PREDICTION

of Extreme Events
in Nature and Society

V. I. KEILIS-BOROK
edited by A. A. Soloviev

Ori Books

Prediction of Extreme Events in Nature and Society

Author: V. I. Keilis-Borok
Editor: A. A. Soloviev

Ori Books is an imprint of Dragonwell Publishing
(www.dragonwellpublishing.com)

ISBN 978-1-940076-45-4
DOI: 10.28935/9781940076447

This edition contains only the original text by the author and
does not include the appendix with reprints

V. I. Keilis-Borok has worked on this book during the past few years of his life. The current text was compiled from the major draft he left behind, completed and edited by one of his closest colleagues, A. A. Soloviev.

The book's target audiences include scientists and professionals interested in learning more about V. I. Keilis-Borok's prediction methods. This book can also be used as a reference source and as a guide for experts and non-experts interested in prediction of extreme events in complex systems.

ABOUT THE AUTHOR

V. I. Keilis-Borok (1921-2013) is one of the founders of computational seismology and mathematical geophysics and a pioneer of advance predictions of extreme events in complex systems, including earthquakes, economic recessions, outcomes of elections, surges of unemployment, and crime waves. His main idea is the consideration of the Earth lithosphere as a complex hierarchical system and earthquakes – as extreme events in it. Among successful advance earthquake predictions made by his group are Irpinia earthquake in Italy (1980), Loma Prieta earthquake in California (1989), Chile earthquake of 2010, Japan earthquake of 2011, and many others.

At the end of the last century V. I. Keilis-Borok began to apply his earthquake prediction approaches on the prediction of extreme events in other complex systems: economic recessions, outcomes of elections, surges of unemployment, and crime waves. His trademark style involved exceptional organizational talent and insight that enabled him to make seemingly impossible connections between different fields of research and different groups of experts, often breaking the barriers between high theory, numerical modeling, and data analysis.

He was the first chairman (in 1964-1979) of the International Working Group on Geophysical Theory and Computers (now – the IUGG Commission on Mathematical Geophysics), and a leader of numerous scientific organizations worldwide. In the last years of his life he held a faculty position at UCLA. His outstanding achievements are recognized by memberships in many academies and international scientific organizations, including National Academy of Sciences of the USA, American Academy of Arts and Sciences, Academy of Sciences of the USSR (now – Russian Academy of Sciences), Pontifical Academy of Sciences, Academia Europea, Royal Astronomical Society, and others, as well as through awards and honors, and by the respect and devotion he inspired in his friends and colleagues.

The book contains an overview of over fifty years of work by his group on developing successful algorithms for prediction of extreme events in nature and society, including predictions of earthquakes, outcomes of elections, and socioeconomic crises.

ABOUT THE EDITOR

A. A. Soloviev belonged to V. I. Keilis-Borok's team from 1976 to 2013 and his studies included application of pattern recognition approaches to solving geophysical problems (in particular, recognition of earthquake prone-areas), magnetic dynamo problem, numerical simulation of block structure dynamics and seismicity etc. He played an important role in achievements by V. I. Keilis-Borok in the field of prediction of extreme events in socioeconomic systems.

A. A. Soloviev is a corresponding member of the Russian Academy of Sciences and participated in several national and international projects focused on prediction of extreme events and seismic hazard estimation. From 1998 to 2017 he has served as the Director of the Institute of Earthquake Prediction Theory and Mathematical Geophysics, Russian Academy of Sciences (MITPAN – IEPT RAS), founded by V. I. Keilis-Borok in 1990 in Moscow. He has collaborated with V. I. Keilis-Borok on this book until October, 2013.

TABLE OF CONTENTS

INTRODUCTION

Today's civilization is highly vulnerable to crises arising from extreme events generated by complex and poorly understood systems. Examples include natural disasters, external and civil wars, terrorist attacks, crime waves, economic downturns, and famines, to name just a few. Yet more subtle effects also threaten modern society, e.g., the inability of democratic systems to produce policies responsive to challenges, such as climate change, global poverty, and resource depletion.

Our capacity to predict the course of events in complex systems is inherently limited. However, there is a new and promising approach to predicting and understanding complex systems that has emerged through the integration of studies in the natural and social sciences and the mathematics of prediction. This book describes and analyzes this approach and its real-world applications. Originally this approach was successfully applied to natural disasters, such as earthquakes. Further developments include algorithmic prediction of electoral fortunes of incumbent parties, economic recessions, surges of unemployment, and outbursts of crimes. This leads to important inferences for averting and responding to impending crises.

Ultimately, improved prediction methods enhance our capacity for understanding the world and for protecting and sustaining our civilization.

Hierarchical complex systems persistently generate extreme events – rare, fast changes that have a strong impact on the system. Depending on connotation they are also known as critical phenomena, disasters, catastrophes, and crises. This book summarizes decades of work on the development and application of algorithmic predictions to extreme natural, socio-economic, and political events.

The studies presented here summarize over 50 years of research by teams from over 20 institutions in 12 different countries, led by V. I. Keilis-Borok and stemming from his original discoveries, open a reasonable hope to break the current stalemate in disasters' prediction, in application to multiple extreme and catastrophic events in nature and society.

BASIC SCIENCE FOR THE SURVIVAL OF HUMANITY IN THE THIRD WORLD WAR

M. Alfimov, R. Corell, V. Courtillot, V. Fortov, M. Intriligator, and V. Keilis-Borok

This article, initiated by V. Keilis-Borok, was originally published in Russian in a national newspaper "Kommersant Daily", November 29, 1997

At the time of this publication, M. Alfimov was the Chairman of the Russian Foundation for Basic Research; R. Corell was Assistant Director of the US National Science Foundation; V. Courtillot was Director of the Institute of the Physics of the Earth, Paris; V. Fortov was the Vice President of the Russian Academy of Sciences; M. Intriligator was Professor of Economics, Political Science, and Policy Studies at the University of California, Los Angeles; V. Keilis-Borok was the Director of the International Institute of Earthquake Prediction Theory and Mathematical Geophysics, Russian Academy of Sciences.

It is commonly recognized that the very survival of our civilization is threatened by dangers as great as the danger of nuclear war. They include the destabilization of the environment, the economy, and the social order due to man-made and natural disasters; the self-destruction of megacities; pandemic terrorism; the loss of control over telecommunications systems; and the depletion of natural resources. These dangers keep growing despite huge sums of money being spent on their containment. Meanwhile, modern basic science has created new revolutionary possibilities to break this stalemate, and humankind cannot afford to ignore them. Their realization requires the same spirit of responsibility, excellence, and urgency as drove the defense-oriented projects in World War II.

Hundreds of billions of US dollars are being spent worldwide in an effort to contain these dangers by the massive application of existing technologies. These efforts may prevent a part of the potential damage, but, on the whole, they are in a stalemate: the factors destabilizing our civilization prevail, the scale of possible catastrophes is rapidly growing, and the solution is generally believed to lie in spending ever more money.

Currently, a massive release of radioactivity from a nuclear waste disposal site, an earthquake in the middle of a major city, a large-scale outburst of violence, and any one of a formidable array of other quite possible disasters could cause millions of casualties, render a large part of our world uninhabitable, trigger a global economic depression, and even trigger a war in a "hot" region. In addition, each country has become vulnerable to developments in other parts of the global village that are outside its control. Meanwhile, looming on the horizon is the critical problem of sustainable development on a longer time scale.

We know from history and common sense that basic research is pivotal to breaking such a stalemate. Indeed, since ancient times basic science has time and again rescued humanity from major threats and sustained its development by creating a springboard to entirely new technologies and thereby providing "new solutions to old problems." Among recent examples are antibiotics, transistors and integrated circuits, synthetic fibers, and the green revolution, to name just a few. Thus, the hope and responsibility to break this stalemate rest not with the Treasury but rather with the Academia. Still the Treasury is more popular, as in the French proverb: "Nobody is satisfied with his wealth, everybody is satisfied with his wisdom."

Frontier research of the last years continues this tradition, discovering new possibilities to cope with many of the present dangers. To be specific, let us mention some of them.

In the area of disaster reduction, they include neutralization of the time bomb contained in radioactive wastes; the prediction of natural disasters; geo-engineering stabilization of megacities; the control of the traffic in chemical explosives; the control of telecommunication networks; and the prediction of social, economic, and political crises.

In the area of sustainable development, they include the discovery of new mineral deposits; the creation of new materials and sources of energy; the development of new forms of transportation; and the processing of large waste disposals.

Now is the time to launch a series of "superprojects" to explore these kinds of possibilities. Inevitably, they will be too unorthodox to be supported via the usual procedures. A classical example is the research that led to the development of radar in England prior to World War II. Funding of this research was at first objected to in that it detracted from the limited funds for "more immediate" goals, such as the improvement of barrage balloon systems. Funding was granted due to the interference of Winston Churchill, and radar ended up saving England in the first year of the war. Another, but this time tragic example, was the initiative of Leonid Kantorovich, later to be a Nobel Laureate in Economics. In 1941 he suggested that the Soviet authorities optimize the routing of railroad trains using the technique of linear programming that he had just invented. This proposal was rejected as an arcane irrelevancy, yet it would have increased by many times the capability of the rail system (during the war!).

Both examples illustrate the fact that research of the kind discussed here, by its very nature, cannot be established within existing programs. For example, the development of radar could not be supported by the program of barrage balloon development. On the other hand, such research would not represent much of an economic burden since it would require only a fraction of the funds already allocated to similar goals. Moreover, military R&D facilities could be engaged at the subsequent development stage, rather than undergoing, as they are now doing, costly, slow, and piecemeal conversion. As a matter of fact, one of the difficulties of starting such a project may be just the opposite of the lack of funds: the cost to start will be in the millions rather than the billions of dollars and, accordingly, the project will not be considered at the appropriate level of authority.

Research of such significance, urgency, and difficulty would require wide international collaboration, engaging the top scientists and research facilities of many nations. Such collaboration will make feasible the goals that no country can accomplish alone even if it has large resources of its own. At the very least, such collaboration will save each country considerable time and funds in R&D. The major opportunities in this respect lie in collaboration with Russia. Despite its present perils, Russian science retains immense resources, including active research groups working at the state of the art level; unique experimental facilities; and, in a broader view, leading scientific schools with strong links between abstract and experimental research.

To establish such projects would require a rare commodity: bold, innovative, and responsible initiatives. A major difficulty lies in the self-destructive dwindling of interest in basic research after each major power lost its archenemy. Another difficulty, coming from the side of the scientific community, is its fatalistic lack of persistence in its own initiatives.

Governments are aware of the stalemate, and only basic science may break it. How is it possible to make a connection between them? In peacetime there is neither a mechanism nor a tradition by which the most important and revolutionary proposals could be developed and decided upon with the speed and expertise commensurate with their urgency.

We are suggesting a simple straightforward mechanism to break the stalemate by challenging the scientific community to come up with specific initiatives. The first step would be to invite scientists to submit brief (say, five page) pre-proposals based on previous studies. Second, a panel of outstanding scientists will review these pre-proposals and award grants to work out detailed proposals. Such grants are necessary since developing detailed proposals is bound to require a number of meetings, considerable work on the text, etc. Some deviations from the usual review process will probably be necessary. For example, recent achievements of the authors of proposals should be given a larger than usual weight and the authors should be invited to discuss objections to their proposals. This will help to ensure that the most outstanding ideas are not rejected simply because they are too unusual. There are many such ideas which scientists have given up on due to the difficulty of muddling through channels.

These proposals will be the final product of this venture. The history of basic research gives us assurance that some of them will be sufficiently compelling to be funded by an appropriate source. Depending on the nature of a proposal, this source could be a government agency, an international organization, a consortium of private foundations, etc. This venture will be a major success if even one of the proposals would generate its own support; that is if the decision makers could not afford to turn it down.

Similar procedures have been successfully applied on a more limited scale. For example, the U.S. National Science Foundation had developed in this way cooperative projects with South American countries. A venture of this type is also planned by the Russian Foundation for Fundamental Research. Purdue University, in the U.S., has already established a program of grants "to support the preparation of competitive proposals", where grants are awarded on the basis of a two-page outline.

We now answer some possible objections of the skeptics.

First objection: "As Vannevar Bush put it, 'Scientific progress on a broad term results from the free play of free intellects, working on subjects of their own choice, in the manner dictated by their curiosity.' Accordingly, frontier basic research is, by definition, unpredictable, and the best way to get from it major practical results would be not to focus on them in advance." Our reply is that we speak here on such areas of research, which already show a reasonable hope of addressing critically important practical problems.

Another objection: "Governments always respect and support basic science but there are, regretfully, more urgent needs.". Our reply is that we are addressing exactly such needs.

Another: "Maybe, in these lean times, one should support only the projects, which may lead to important applications?" Our reply is that this would be as impossible as supporting not an apple tree but only the apples: basic research requires a professional environment. Moreover this would be short-sighted, since one may then miss the most important possibilities that are yet unnoticed.

Another: "Recent studies of science policy and priorities have not recommended such projects." Our reply is that what we suggest is complementary to these comprehensive studies, that involve more focused goals and a shorter time scale. Metaphorically, we are proposing a sort of commando operation.

Another: "You are exaggerating the dangers." Our reply is that the dangers we are referring to are widely recognized and are already the subject of large-scale funding. What is not recognized, however, is that they cannot be contained without basic research.

Another: "Why don't the projects of such kind appear all the time?" Our reply is that, in fact, they do appear, mainly in the private sector, such as in biotechnology, where fundamental discoveries can be picked up by industry within a few weeks.

However, in many critically important areas the responsibility rests on the agencies, operating on a *cost-plus* basis, through big business formed around them. Obviously, they have lesser built-in safeguards against reliance on costly and inadequate traditional approaches, with the resulting stalemate explained by the lack of funding and not by the necessity of frontier research and new approaches.

An objection of quite a different kind: *"What if the scientific community would not come up with such ideas?"* We cannot imagine this, and it is certainly important to try.

We are suggesting here a first step in merging the international resources of basic science in a new type of war effort: in some 30 Manhattan projects, not in a shooting war or a cold war but rather in joint defense of survival and sustainable development of our civilization.

PREDICTIVE UNDERSTANDING OF EXTREME EVENTS

By V.I. Keilis-Borok and A. Gabrielov

This lecture, given by A. Gabrielov at the symposium "Four paradigms in predicting extremes: Legacy of Vladimir I. Keilis-Borok", (part of the 30th IUGG Conference on Mathematical Geophysics, a series founded by V. Keilis-Borok) in Merida, Yucatan, Mexico, 2-4 June 2014 was based on the talks given by V. Keilis-Borok. Thus, we assign dual authorship to this presentation, transcribed from A. Gabrielov's talk.

I.   Lithosphere as a complex system.

In classical geology, earthquakes have always been considered as simple and isolated events, where a crack forms in the Earth's crust and propagates across a flat surface of a tectonic fault. Foreshocks and aftershocks have been long known, but no one has ever considered the possibility of interactions between earthquakes, or attempted to view the whole structure of the lithosphere as a single complex system. These concepts, introduced by V. Keilis-Borok in the 1960s, have revolutionized our understanding of the earthquakes.

There are several reasons why the lithosphere can indeed be viewed as a complex system. Geometrically, it is a **hierarchical structure** characteristic for classical complex systems. A typical tectonic map always appears hierarchical, with major, secondary and tertiary elements – it is essentially a mess of blocks and faults. Classical geology used to study each fault separately. There used to be classifications of faults, but even in these classifications, each of their properties essentially is based on the properties of single faults.

The idea of hierarchical structure of the lithosphere comes from geomorphology (the work of E. Rantsman) and plate tectonics. In plate tectonics, the dynamics of the lithosphere is reflected in the movements of large, global-scale plates, but there are structural blocks within these plates that can be recognized on all scales. The entire lithosphere is divided into a hierarchy of blocks, from ~ 10 tectonic plates, to ~10^{25} grains of rocks. These blocks are separated by boundary zones, 10-100 times thinner. They are called *fault zones* (starting with the boundaries between the plates) that include *faults, sliding surfaces,* and, finally, *interfaces* between the grains of rocks.

Except for a few lowest ranks, these boundary zones have similar hierarchical structure with much denser fragmentation. Most densely fractured mosaic *"nodes"* are formed around intersections and junctions of these zones; this is due to collision of the blocks' corners. For brevity, a system of boundary zones and nodes is called "fault network".

Along some boundaries, movements can be slow or hidden. In these cases, geologists don't count these boundaries as faults, but they are still seen on the maps and aerial/space photos. Such boundaries with slow or hidden movements are called lineaments.

The structural blocks distinguished by geomorphology are three-dimensional. However, since their horizontal scale is very extensive compared to the thickness of the lithosphere, we consider these blocks two-dimensional, and faults as one-dimensional structures, or lines. This is a simplification, of course. The faults can be kilometers wide, and the intersections of faults are not merely geometrical points but have their own size. So, it is important to understand that even the zero-dimensional fault intersection has its own scale. This scale can be from plate boundaries to grains of rock.

In separate observations, laboratory experiments modeled earthquakes on rocks samples, and the intersections of faults in such experiments were indeed grains of rock. Thus, the global picture includes all kinds of scales.

There are some aspects of self-similarity in this structure. It is not exactly scale-invariant (nothing is, in the real world). There are ideal scale-invariant structures in mathematics, but there is a certain range of scale invariance that is observed, and thus the parameters, such as fractal dimensions, can be calculated. All this already is complex enough to be considered a complex system, because the blocks are moving, their relative movements are observed as displacements along the faults, and there are nodes where movements in different directions are observed in most complicated and least understood ways. These nodes are the nucleation points of earthquakes.

This is the geometric part. Next comes the dynamic part – **the instability of the strength and stress fields**.

It is recognized that the stress in the lithosphere is not homogeneous, there are places where stress concentrates and other places where stress is released. Stress and strength can be very variable across space. In some nodes, strength decreases, and in some nodes a lock-in effect occurs: a fault becomes locked and stress accumulates until it becomes higher than the strength of the fault, but the fault does not move. This is the most dangerous situation that may be catastrophic.

A fault network is a stockpile of instability. This instability includes both "physical" instability, originated at micro level, and "geometric" instability, controlled by the geometry of the fault network at the macro level.

Movement of the blocks is controlled by a stress - strain field. The stress, and particularly the strength, are in turn controlled by a multitude of mechanisms that include filtration of fluids, stress corrosion, fracturing, buckling, petrochemical and phase transitions, etc. Each process may cause up to 10^5 drop of strength. None of them is consistently dominant: even a grain of rock can act as viscoelastic element. Other elements can include aggregates of crystals, source/absorber of fluid, volume, heat, etc. Altogether these mechanisms turn the lithosphere into a nonlinear complex system.

A dramatic drop of strength in a fault would result in a strong earthquake.

In seismically active regions a large part of the movement is realized through earthquakes. There are few places within such regions where the movement is silent, so no earthquakes are observed, but as a rule the movement in the lithosphere is not continuous but results in bigger or smaller drop.

One striking example is the area west of Vancouver, where tectonic plates are moving in respect to each other, but the actual fault is not seen, which means that this fault is locked. It is expected that this area at some point may produce a major earthquake (with the magnitude 9 or more) that would have a catastrophic effect on the whole region.

Now we come to the problem of earthquake prediction, which is in itself contradictory and not as simple as one would expect.

II. Earthquake prediction

There are multitudes of scales in space, time, and energy, in which earthquakes happen. Because of this, the task should not be poised as an attempt to predict an earthquake of a particular magnitude in a particular place. Instead, earthquake prediction should be a process that includes several stages and scales in time and space. Thus, the overall problem of earthquake prediction is posed as the step by step reduction of the space-time window where a strong earthquake may occur.

There are **five stages of prediction**:

1) Background prediction (100 years and more, to identify the places where strong earthquakes may occur);

2) Long-term prediction (10 years or so);

3) Intermediate-term prediction (a few years);

4) Short-term prediction (less than a year);

5) Immediate prediction (days to minutes).

V. Keilis-Borok's earthquake prediction program was originally concentrated on the background predictions and identification of earthquake-prone areas. Much of this work was done in collaboration with I. M. Gelfand, F. Press, L. Knopoff, and E. Rantsman. As the next step, he moved to intermediate-term predictions. With shorter time scales, the predictions become less and less reliable. Short-term and immediate predictions are still out of reach for us, except that, when an earthquake does happen, there are now some systems that recognize the first incoming seismic waves and are able to make some adjustments to neutralize or diminish the next incoming wave. But this is more of an engineering task.

V. Keilis-Borok's main tool was pattern recognition. This tool was pretty controversial in the field. Many times he had to answer the questions: "How can you predict something when you don't have an adequate model?" or "Why don't you derive from the first principles?"

He always responded with a quotation from A. Kolmogorov:
"It became clear for me that it is unrealistic to have a hope for the creation of a pure theory [of the turbulent flows of fluids and gases] closed in itself. Due to the absence of such a theory we have to rely upon the hypotheses obtained by processing of the experimental data…"

In other words, there is no first principle in this case, no differential equations that would describe the principle.

Kolmogorov's quotation referred to the theory of turbulence, where his famous scaling law was derived from half-empirical manipulations with dimensions. His law became one of his most famous results in hydrodynamics. In fact, complete theory of turbulence still doesn't exist—and lithosphere is far more complex than fluid dynamics. But one important similarity between these two systems is that in both systems there is no complete theory that would derive behavior from first principles.

So, what do we mean when we say "prediction"?

It is surprising how many different points of view people have when they talk about earthquake predictions. V. Keilis-Borok's view was very specific, and probably unusual:

First, it is important to define the target: **what do we want to predict?** V. Keilis-Borok's goal was to predict rare extreme events. A predictor in this case would be a discrete sequence of alarms. This is opposite to another very popular point of view of probabilistic prediction. For example, seismic risk maps, regularly published by the US Geological Survey, contain probabilities with which strong earthquake may happen in certain areas. This is an entirely different approach to predictions from V. Keilis-Borok's method. Yes, it is nice to have a probability, which is a real number, but in most cases we don't have enough information to generate a real number like this. Thus, a prediction in this case cannot be a subject of classical statistics.

Pattern recognition of rare extreme events is a unique tool that gives a natural framework for integration of modeling, theory, and extensive data analysis.

The next point, introduced by V. Keilis-Borok and still not properly understood, concerns **premonitory phenomena** — something we observe before a strong earthquake happens. There are two kinds of premonitory phenomena. Some are "perpetrators", the events that start happening and eventually cause a strong earthquake to occur. However, immediate causes of a strong earthquake are often very hard to identify. The other kinds of premonitory phenomena are termed "witnesses". Witnesses don't cause extreme events, but they are by-products of a destabilized system that can be observed. Thus, "witnesses" might predict not an extreme event per se, but the destabilization of the system which makes it ripe for an extreme event.

One example is the proverbial "straws in the wind" that precede a hurricane.

The need for holistic approach.

It is not possible to understand a complex system by breaking it apart. Holistic approach, from the whole to details, opens a possibility to overcome the complexity itself and the chronic imperfection of the data.

Figure P1. The need for holistic approach: a fracture in a rock cannot be seen in a close-up view (left), but becomes far more evident at a distance (right).

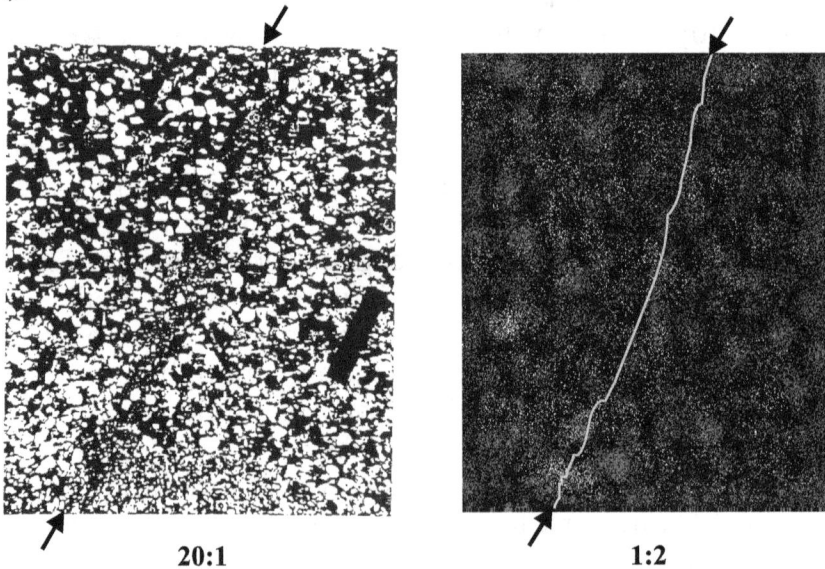

20:1   1:2

Fig. P1 is an illustration of a geological experiment done a while ago. It shows images of a rock at two different magnifications. If you look at the rock too closely (left) you don't see anything but a chaotic mess of grains. But if you zoom out and look at the same rock on a smaller scale (right), you see a fracture, which is completely lost in the higher magnification view on the left. This was one of the main V. Keilis-Borok's themes – if you want to study the system as a whole you need to get spacial and temporal averages, smoothing things to understand large scale structures that will be lost in a too-detailed view. This smoothing is essential in prediction of strong earthquakes.

Quoting the famous American physicist M. Gell-Mann:

"If the parts of a complex system or the various aspects of a complex situation, all defined in advance, are studied carefully by experts on those parts or aspects, and the results of their work are pooled, an adequate description of the whole system or situation does not usually emerge... The reason, of course, is that these parts or aspects are typically entangled with one another... We have to supplement the partial studies with a transdisciplinary crude look at the whole."

In other words, if you study separate aspects of a complex system and then combine them, you may lose the properties of the system

Figure P2: Attacking herd of elephants as an example of a complex system behavior.

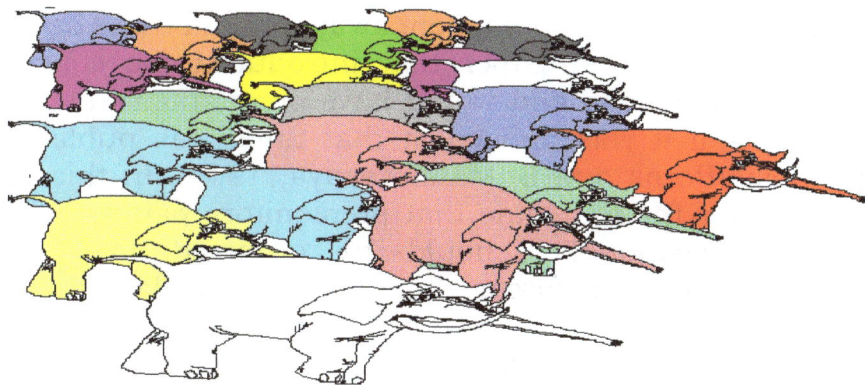

as a whole, which may only appear when elements interact with each other.

This kind of a behavior can be well illustrated by an attacking herd of elephants (Fig. P2). If you look at a single elephant, you would see him extend his trunk and raise his ears, but you may not recognize that the whole herd is about to attack. Only looking at the herd as a whole makes it obvious.

One example of why large-scale observations are essential in prediction of earthquakes is the Parkfield experiment, in the early 1980's. The experiment involved detailed observation of seismic activity in Parkfield, CA. In this area, which includes a segment of the St. Andreas fault, earthquakes have been occurring nearly every 21 years with some exceptions. The project involved installing a lot of monitoring devices to measure tectonic stress, microseismicity, fluid levels, and everything else they were able to measure, in a small area where the next Parkfield earthquake was expected to happen. Eventually the earthquake occurred 20 years later than expected, and none of the devices were able to detect anything preceding this earthquake. But during this same period of time strong earthquakes happened in several other places relatively nearby – Northridge, Loma Prieta, and others. The participants of this project ignored all these other earthquakes, waiting for another Parkfield. In reality, they should have been looking at a much larger area.

Another example concerns an attempt to predict an earthquake in Tokyo. If one occurred, it would have been very damaging for Tokyo. To predict it, a project was initiated and a lot of devices were installed to monitor different parameters in the immediate Tokyo area, but then the Kobe earthquake happened in a region where nobody expected it to happen, away from all these devices.

V. Keilis-Borok's first paper that introduced the concept of premonitory phenomena and would have pre-empted this kind of failed attempts if it was noticed at that time, was published in 1964 (V. Keilis-Borok and L. Malinovskaya, 1964). In this paper, they reported for the first time that premonitory patterns leading to a strong earthquake should be observed in an area approximately 10 times larger than the earthquake itself. It is definitely useless to look exactly at the place where the earthquake is going to happen to predict an earthquake. Historically, all efforts to do it have failed. The first stage of earthquake prediction following this discovery was originally started by V. Keilis-Borok in collaboration with the prominent Russian mathematician I. M. Gelfand, the American scientists L. Knopoff and F. Press, and a Russian geomorphologist E. Ya. Rantsman (Fig. P3). The figure shows two maps: on the right is the map of the blocks and lineaments defined by Rantsman and her group using a combination of geological, geomorphological, and tectonic data. On the left is the map of the major known tectonic faults. The points on the left-hand map, indicating the sites of the strong earthquakes that historically occurred in these areas, corresponded to the nodes identified in V. Keilis-Borok's analysis as the sites which could serve as the points of the nucleation of strong earthquakes. The goal of this work was to use historical data and a list of properties of the nodes to distinguish between the nodes where the strong earthquake may or may not nucleate. Using the two maps gives an overall similar results, but the more detailed map on the right reveals more nodes, some of them dangerous but not seen in the tectonic fault map.

Figure P4. Strong earthquakes nucleate in some "dangerous" structures (D-nodes) (from Gelfand et al, 1976).

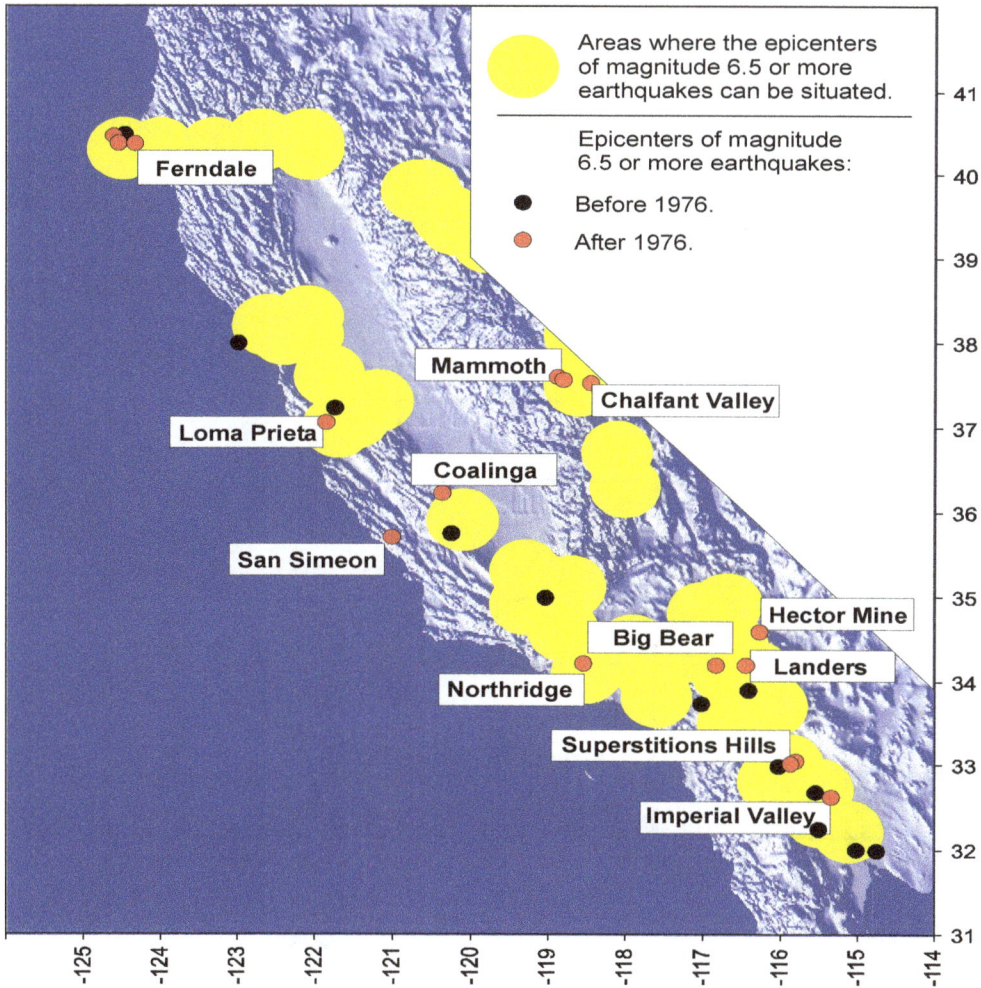

In fact, some nodes identified by this work later became the sites of strong earthquakes – for example, Northridge, which nobody expected to become the earthquake site. In California everyone was looking at another fault, and construction codes were not prepared for the Northridge earthquake. After Northridge, many other places were identified, which were not on the active faults, but were similar to the known places on the active faults. Many areas that were originally missed as potential earthquake sites were mapped – e.g., in downtown Los Angeles.

V. Keilis-Borok's group used pattern recognition to identify such "dangerous" nodes (Fig. P4). The identified nodes are shown in yellow, and black dots denote the sites of earthquakes that happened before 1976 and were used in the analysis. Red dots show the earthquakes that happened after 1976. Several of them occurred in the areas that were not previously considered dangerous.

Fig. P5 shows what an algorithm for earthquake prediction should look like in V. Keilis-Borok's approach. For simplicity, this algorithm does not consider space, only time. Vertical dashed line

Figure P5. A model algorithm for an earthquake prediction, in V. Keilis-Borok's approach.

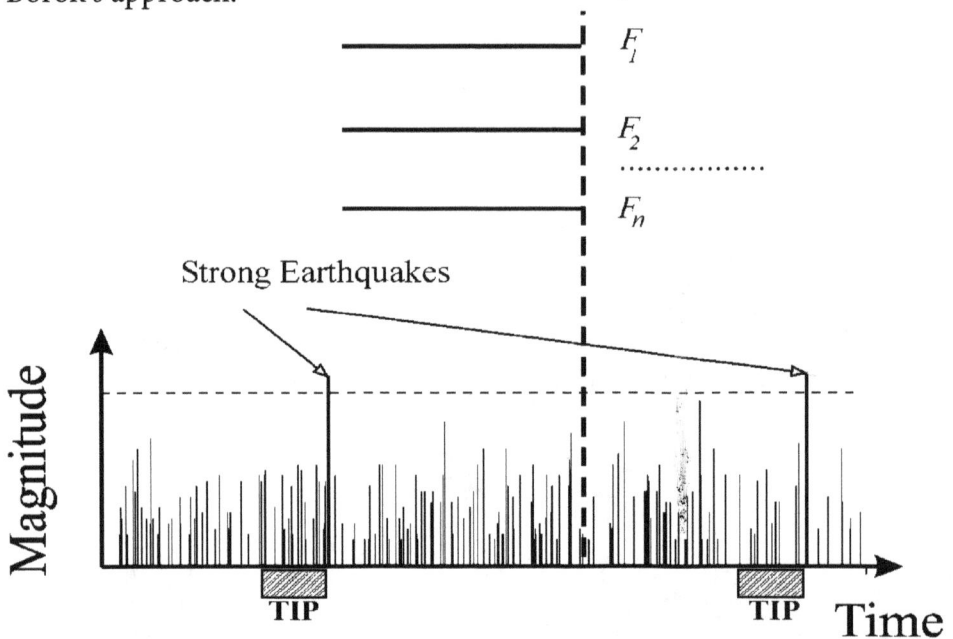

shows a given time moment. From this moment, we take a time window backwards (defined as an observation window) and compute some functionals based on observations within this window—usually some integral patterns of seismicity (e.g., total seismic activity, clustering, etc.). We next associate several numbers to our time moment and look at the historical statistics of those numbers to compare their frequency just before a strong earthquake (dangerous period), excluding the period just after (unstable period), but also considering longer periods before and after. When the actual observed values reach the extreme, alarm is declared (TIP- **t**ime of **i**ncreased **p**robability of a strong earthquake).

The main requirements for this kind of algorithm are that (i) it must be objective, (ii) enough data must be available to compute those functionals, and that (iii) the data should be homogeneous. The real predictions also consider space were the earthquake will happen.

The diagram in Fig. P5 shows two strong earthquakes and two TIPs.

Possible outcomes of predictions range in time and space, from false alarms and failures to correct alarms and successful predictions. Good and bad prediction methods could be distinguished by comparing prediction to real data.

So, how do we choose the functionals?
The choice is guided by **four paradigms**:
1. Basic types of premonitory phenomena. Premonitory phenomena in V. Keilis-Borok's studies consisted of all the weak and intermediate seismicity data in the area. Seismicity data were the most reliable and uniform data available at the time. Other data (e.g. GPS) did not exist back then, but can be used in future predictions.

2. Long - range correlations in the fault system dynamics. Premonitory phenomena are formed not only in vicinity of the incipient source but within a much larger area. Overall, one needs to look in the area at least 10 times larger than the earthquake area.

3. Partial similarity of premonitory phenomena in the diverse conditions. All the functionals are scaled, so that if you change the premonitory phenomena you can directly compare data in the areas with different levels of seismicity. The same functionals can apply to a variety of phenomena, from small rock bursts in mines, and induced seismicity in basins, as well as the largest earthquakes magnitude (M) from 4.5 to 8+ worldwide – and, possibly, to starquakes. The energy can range from erg to 10^{26} erg, and possibly to 10^{41} erg.

4. The dual nature of premonitory phenomena. Some of the premonitory phenomena are "universal", common for complex

Figure P6. An extreme event is preceded by premonitory changes of the background activity (its deviations from long-term average).

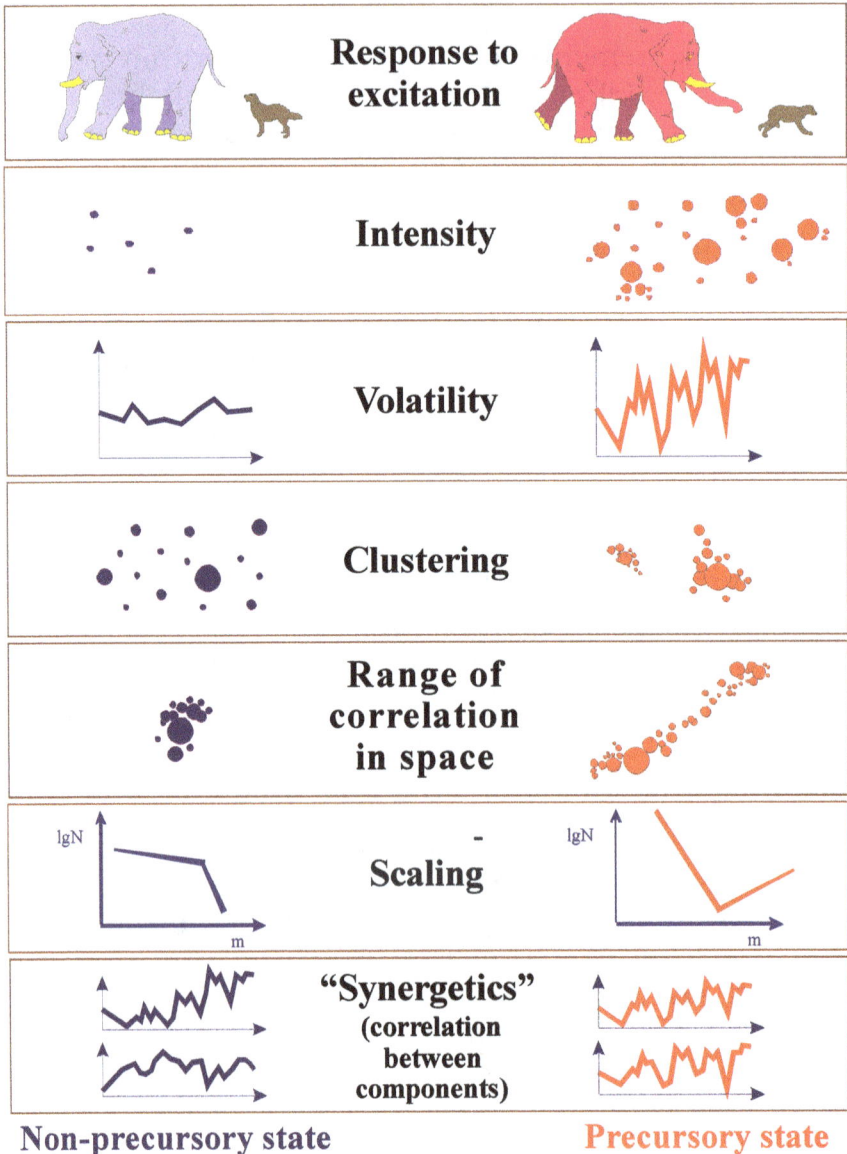

	Response to excitation	
	Intensity	
	Volatility	
	Clustering	
	Range of correlation in space	
	Scaling	
	"Synergetics" (correlation between components)	
Non-precursory state		**Precursory state**

nonlinear systems of different origin; others are Earth-specific. According to the general theory of complex nonlinear systems, these systems can exhibit abrupt qualitative changes — bifurcations, phase transitions, and loss of stability. Some of these changes can serve as premonitory phenomena for critical transitions in these systems, such as earthquakes. Some of these events are tied to particular properties of the Earth and appear only in very specific settings of tectonic fault systems. The functionals should take into account these specifics.

Common premonitory patterns.

In defining the premonitory patterns, we are looking at the long-term activity in this system, and when the statistics of this activity becomes extreme, this activity becomes a premonitory pattern for an extreme event (e.g., an earthquake).

Figure. P7. Successful advance prediction of Loma Prieta earthquake in 1989, M=7.1 Light gray – area of investigation; Red - alarm reduced by MSc; Dark gray – alarm by M8; Asterisk – Loma Prieta epicenter

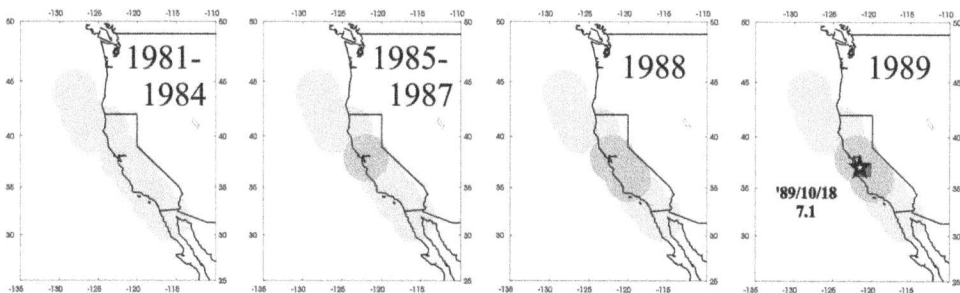

The following changes can serve as premonitory patterns (illustrated in Fig. P6):

1. Response to excitation. In normal state, a system should not show a strong disturbance in response to mild excitation. If it shows a strong reaction, this indicates that the system is unstable.

2. Intensity increase. Activity in an area becomes less uniform, concentrating in some places (not necessarily in the place where an earthquake will happen).

3. Volatility. The amplitude of activity increases.

4. Clustering. Spacial and temporal distribution of events becomes less uniform, concentrating in some places (not necessarily where the earthquake happens)

5. Range of correlation in space. When a strong earthquake approaches, moderate earthquakes tend to appear close in time at faraway places in space.

6. Scaling law. The quiet-time distribution of earthquakes is dominated by smaller ones, the dangerous times are dominated by stronger ones.

7. Correlation between components. This is not properly observed with earthquakes, but is observed in other systems (e,.g, financial markets), thus, hypothetically, this can also serve as a precursor.

Figure P8. Frontiers of similarity.

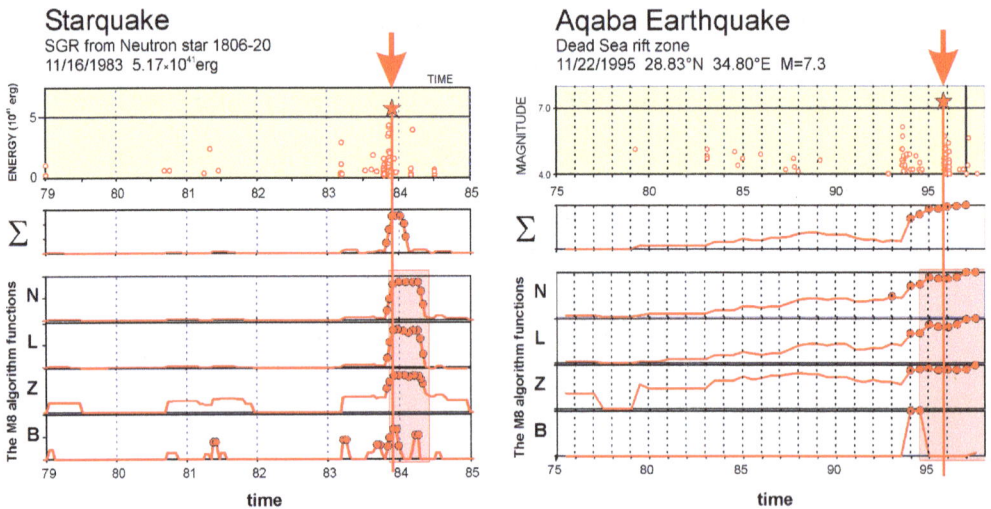

Advance predictions.

Using these approaches, V. Keilis-Borok's group has developed the algorithm M8 (for prediction of earthquakes with the magnitude of 8 and more). One of the first predictions using this algorithm was the Loma Prieta earthquake that occurred in California in 1989. This algorithm was successfully used for decades of earthquake predictions that are still ongoing right now.

The prediction was made 5 years ahead (Fig. P7). In 1981-1984 the dangerous zone was identified (gray in the leftmost panel in Fig. P7). In 1985-1987 one area in that zone became dangerous, and in 1988 two adjacent areas also showed increase seismic activity (dark gray circles in middle panels). The 1989 earthquake happened within those areas. The results of this prediction were reported in a meeting between the Russian President M. Gorbachev and the US President R. Reagan in 1988. After the earthquake happened, V. Keilis-Borok's team from Russia was urgently sent to California.

The same algorithm M8 was able to predict major busts from neutron star and an earthquake in a Dead Sea rift zone, which has fairly low seismicity (Fig. P8).

III.   Socioeconomic predictions

The same approaches can be applied to predictions in socioeconomic systems. In each case, the socioeconomic system in question is considered a complex system, and the functionals are chosen by experts in the appropriate areas as the parameters that reflect the behavior of this particular system. Successful socioeconomic predictions by V. Keilis-Borok's group, outlined later in this book, include predictions of outcomes of US elections, economic recessions, and surges of unemployment and crime (see Fig. P9 and chapters 3-6 of the current book).

Predictions of the US presidential elections.
In collaboration with A. Lichtman, V. Keilis-Borok developed a successful algorithm for predictions of the US presidential elections. This prediction was based on thirteen socio-economic and political factors, later termed "thirteen keys to presidency" or "thirteen keys to the White House" (Chapter 3, section 3.3). Victory of the challenging party is predicted when 6 or more factors are in its favor. Otherwise victory of incumbent party is predicted.

Notably, this algorithm predicted the victory of Donald Trump in 2016, months in advance of the election.

IV. Using the limited accuracy predictions for disaster preparedness.

"Of course, things are complicated....But in the end every situation can be reduced to a simple question: Do we act or not? If yes, in what way?"

E. Burdick

In using predictions for disaster preparedness, it is important to remember the following:

Basic principle: escalate or de-escalate preparedness measures, according to what and where is predicted and what is the quality of prediction.   Such is the standard practice in preparedness to all disasters, war included.

Diversity of damage: failure of constructions; fires; release of dangerous materials; triggering of floods, avalanches, landslides, tsunami, etc.

Socio-economic impact: disruption of vital services - supply, medical, financial, law enforcement etc.; epidemics; disruptive anxiety of population, profiteering and crime; drop of production and employment; destabilization of financial systems.

Figure P9. Socioeconomic predictions in the US.

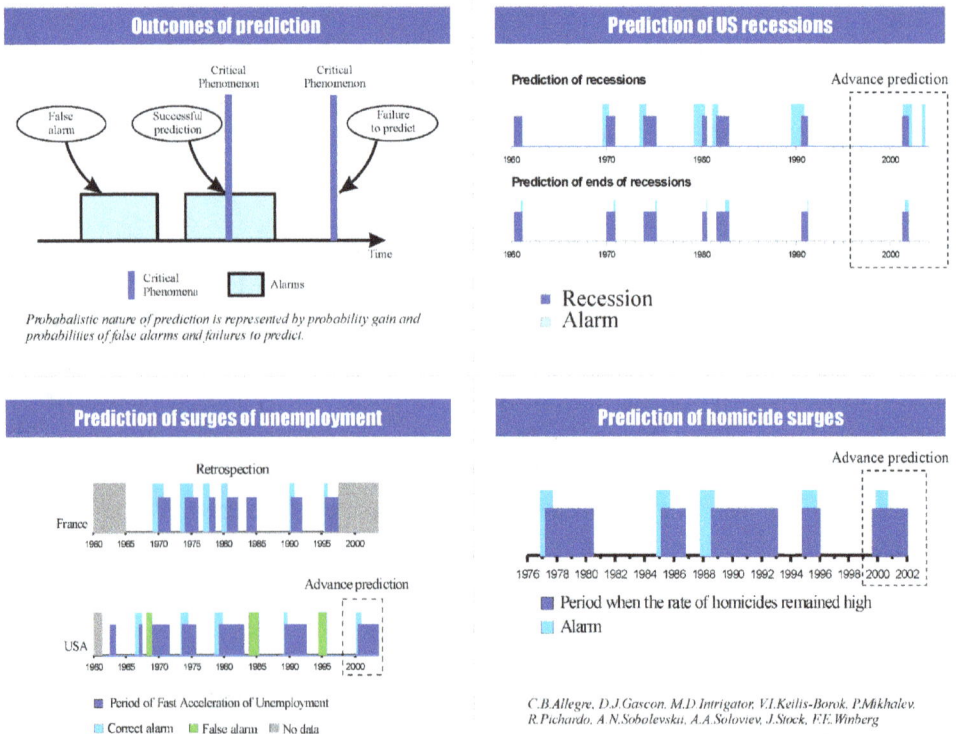

Impact by undue release of predictions: knowledge that an extreme event is about to occur may cause panic, with dramatic consequences on all levels. This was one of the reasons V. Keilis-Borok's group distributed their predictions in a closed way, to a selected group of experts worldwide.

A hierarchy of **diverse safety measures** is required by the diversity of damage from earthquakes. These include:

Background measures: restriction of land use; building codes; insurance and bonds; preparedness of civil defense type; R&D.

Temporary measures, activated in response to a time prediction: enhancement of permanent measures - safety inspections, simulation alarms etc.; partial neutralization of high - risk objects; mobilization of post - disaster services; emergency legislation, up to martial law; evacuation of population etc.

These measures are required in different forms on local, provincial, national and international levels.

Different measures require different lead time, from seconds to years, to be activated; having different cost they can be realistically maintained for different time - periods, from hours to decades; and they have to be spread over different territories - from selected points to large regions.

Different stages of earthquake prediction are useful for damage reduction.

No single measure is sufficient alone. On the other hand, many important measures are inexpensive. The accuracy of prediction should be known, but not necessarily high.

CHAPTER 1

Global Approach to Prediction of Extreme Events in Complex Systems

1.1. Holistic Approach

Natural science had, for many centuries, regarded the Universe as a completely predictable machine. As Pierre Simon de Laplace wrote in 1776, "...if we knew exactly the laws of nature and the situation of the universe at the initial moment, we could predict exactly the situation of the same universe at a succeeding moment." However, at the turn of the 20th century (1905) Jules Henry Poincare discovered, that "... this is not always so. It may happen that small differences in the initial conditions will produce very great ones in the final phenomena. Prediction becomes impossible". This concerns nonlinear (chaotic) systems that contain mutually dependent mechanisms destabilizing their behavior. The complexity is an attribute of such systems. It comprises instability and complex, but not random, behavior patterns – "order in chaos." The complex systems are not predictable with absolute precision. However, after coarse-graining (on a not- too-detailed scale), premonitory phenomena emerge, and a system becomes predictable, up to the limits (Farmer and Sidorowich, 1987; Ma et al., 1990; Kravtsov, 1993; Gell-Mann, 1994; Holland, 1995; Kadanoff, 1976; Crutchfield et al., 1986).

The holistic approach, "from the whole to details", opens a possibility to detect "universal" types of precursors overcoming the complexity itself, and the chronic imperfection of observations as well. Through robust integral description of the complex systems, it is possible to discover their regular behavioral patterns transcending their inherent complexity. This is achieved at an unavoidable price: the accuracy of prediction is limited. The holistic approach that proceeds from the whole to details is opposite to the reductionism approach that proceeds from details to the whole. It is in principle not possible "to understand a complex system by breaking it apart" (Crutchfield et al, 1986). Table 1 compares the holistic approach with the complementary (but not necessarily contradictory) reductionism approach.

Among the regular behavior patterns of complex systems are "premonitory" ones that emerge more frequently as an extreme event approaches. These premonitory patterns make complex systems predictable. The accuracy of these predictions, however, is inevitably limited, due to the systems' complexity and observational errors.

1.2. The Prediction Problem and Premonitory Patterns

If we consider extreme events that concern the complex system as a whole, then the prediction problem can be formulated as follows:

given are time series that describe dynamics of the system up to the current moment of time t and contain potential precursors of an extreme event;

to predict whether an extreme event will or will not occur during the subsequent time period $(t, t + \tau)$; if the answer is "yes", this will be the *"period of alarm."*

As the time goes by, predictions form a discrete sequence of alarms. The possible outcomes of such a prediction are shown in Fig. 1. The actual outcome is determined unambiguously, since the extreme events are identified independently of the prediction. In some problems considered here they are routinely determined independently of our research (e.g. earthquake occurrence or an election result). In other problems they are defined by a by a separate algorithm (e.g. homicide surge).

Such "yes or no" prediction is aimed not at analyzing the whole dynamics of the system, but only at identifying the occurrence of rare extreme events. In a broad field of prediction studies this prediction is different from, and complementary to, the classical

Kolmogoroff–Wiener prediction of continuous functions, and to traditional cause-and-effect analysis.

Table 1. Two complementary approaches to prediction

"REDUCTIONISM" *(from details to the whole)*	"HOLISM" *(from the whole to details)*
Premonitory phenomena preceding an extreme event are formed:	
in the narrow time-space vicinity of the future event	in the wide time-space volume containing the future event
Premonitory phenomena are:	
specific to mechanisms controlling the behavior of the complex system	divided into: – "universal" ones common for many complex systems; – those, depending on the complex system structure; – mechanism-specific ones.
Premonitory phenomena in different complex systems and energy ranges are:	
different	to a considerable extent similar
Constitutive equations are:	
local	non-local
Triggering of extreme events is controlled:	
by the system state in the time-space vicinity of the future event	by the state of the whole system during a rather long period before the future event

The problem includes estimating the predictions' accuracy: the rates of false alarms and failures to predict, and the total duration of alarms in % to the total time, are considered. These characteristics represent the inevitable *probabilistic component* of prediction; they provide for statistical validation of a prediction algorithm and for optimizing preparedness to predicted events (e.g. recessions or crime surges).

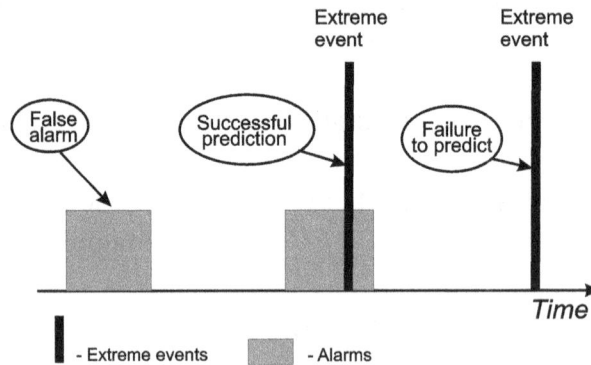

Figure 1. Possible outcomes of prediction

Twofold importance.

Prediction problem is pivotal in two fields.

— *Fundamental understanding of complex systems.* Prediction algorithms quantitatively define phenomena that anticipate extreme events. Such definition is pivotal for fundamental understanding of a complex system where these events occur, including the intertwined mechanisms of system's development and its basic features, e.g. multiple scaling, correlation range, clustering, fragmentation etc. The understanding of complex systems remains a major unsolved problem of modern science, tantamount to transforming our understanding of the natural and human world.

— *Disaster preparedness.* On the practical side, prediction is pivotal for coping with a variety of disasters, commonly recognized as major threats to the survival and sustainability of our civilization (e.g. Keilis-Borok and Sorondo, 2000; Davis et al., 2010; see also materials of G8-UNESCO World Forum on "Education, Innovation and Research: New Partnership for

Sustainable Development", http://g8forum.ictp.it). The reliable advance prediction of extreme events can save lives and contribute to social and economic stability, and to improving the governing of modern societies.

Certain systems exhibit behavior patterns that emerge more frequently as an extreme event draws near. These patterns, called premonitory patterns, signal destabilization of the system and thus an increase in the probability of an extreme event. Premonitory patterns do not necessarily contribute to causing a subsequent extreme event. Both the patterns and the event itself might be parallel manifestations of the same underlying process occurring in the complex system, formed in multiple time-, space-, and energy- scales. For this reason premonitory patterns might emerge in a broad variety of observable fields reflecting complex system dynamics, and in different scales.

Prediction algorithms are designed by analysis of the "learning material" – a sample of past critical events and the time series hypothetically containing premonitory patterns. Typically, for complexity studies we do not have a complete set of fundamental equations that govern dynamics of the system and unambiguously define prediction algorithms. In lieu of such equations "…we have to rely upon the hypotheses obtained by processing of the experimental data" (A. Kolmogorov, 1943 on transition to turbulence). A methodology for designing the prediction algorithms is the *pattern recognition of rare events* that was developed by the school of I. Gelfand for studying rare events of highly complex origin (e.g., Bongard, 1970; Gelfand et al., 1976; Keilis-Borok and Press, 1980; Keilis-Borok and Lichtman, 1993; Press and Allen, 1995; Keilis-Borok et al., 2000, 2003, 2005, 2008).

Algorithm design comprises three following steps.

1. *Detecting premonitory patterns.* Each time series considered is robustly described by the functionals $F_k(t)$, $k = 1, 2, …,$ capturing hypothetical patterns. Hypotheses on what these patterns may be are provided by "universal" modeling of complex systems, modeling of system-specific processes, exploratory data analysis, and practical experience, even if it may be intuitive. Pattern recognition of rare events is an efficient common framework for formulating and testing such hypotheses, their diversity notwithstanding. With a few exceptions the functionals

are defined in sliding time windows; the value of a functional is attributed to the end of the window.

2. *Discretization.* Emergence of a premonitory pattern is defined by the condition $F_k(t) \geq C_k$. The threshold C_k is chosen is such a way that premonitory pattern emerges on one side of the threshold more frequently than on another side. That threshold is usually defined as a certain percentile of the functional F_k. In such robust representation of the data, pattern recognition is akin to exploratory data analysis developed by J. Tukey (1977).

3. *Formulating an algorithm.* Prediction algorithm will trigger an alarm when certain combination of premonitory patterns emerges. This combination is determined by further application of pattern recognition procedure (Keilis-Borok and Lichtman, 1993; Press and Allen, 1995; Keilis-Borok and Soloviev, 2003).

1.3. Algorithm Reliability. Error Diagrams

Estimating reliability of an algorithm is necessary, since an algorithm inevitably includes many adjustable elements, from selecting the data used for prediction and definition of prediction targets, to the values of numerical parameters. In lieu of the closed theory a priory determining all these elements they have to be adjusted retrospectively, by "predicting" the past extreme events. The application of the methodology to known events creates the danger of self-deceptive data-fitting: As J. Von Neumann put it *"with four exponents I can fit an elephant"*. The proper validation of the prediction algorithms requires three consecutive tests:

— *sensitivity analysis*: varying adjustable elements of an algorithm;

– *out of sample analysis*: applying an algorithm to past data that has not been used in the algorithm's development;

— *predicting in advance* – the only decisive test of a prediction algorithm.

Such tests take a lion share of data analysis (Gelfand et al., 1976; Gabrielov et al., 2000b; Zaliapin et al., 2003b; Keilis-Borok and Soloviev, 2003). A prediction algorithm makes sense only if its performance is (i) sufficiently better than a random guess, and (ii) not too sensitive to variation of adjustable elements. Error diagrams described below show whether these conditions are satisfied.

Definition. Error diagrams show three major characteristics of prediction's accuracy. Consider an algorithm applied during the time period *T*. During the test *N* extreme events have occurred there, and N_m of them have been missed by alarms. Altogether A alarms have been declared and A_f of them happened to be false. Total duration of alarms is D.

Performance of an algorithm is characterized by three dimensionless parameters: the relative duration of alarms, $\tau = D/T$; the rate of failures to predict, $n = N_m/N$; and the rate of false alarms, $f = A_f/A$. These three parameters are necessary in any test of prediction algorithm regardless of a particular methodology. They are juxtaposed on the error diagrams schematically illustrated in Fig. 2. Also called "Molchan diagrams" in earthquake prediction studies, they are used for validation and optimization of earthquake prediction algorithms and for joint optimization of prediction and preparedness (Molchan, 1990, 1991, 1994, 1997, 2003). In many applications parameter *f* is not yet considered. In early applications they are called ROC diagrams for "Relative Operating Characteristics" (e.g., Mason, 2003).

1.4 Prediction and Disaster Preparedness

What, if any, preparedness actions should be undertaken in response to a prediction, given its inherently limited accuracy? Methodology assisting decision-makers in choosing optimal response to prediction has been developed in Kantorovich et al. (1974) and Molchan (1991, 1997, 2003). Its numerous applications are described by Keilis-Borok et al. (2004), Davis et al. (2007, 2010), and Molchan and Keilis-Borok (2008).

Earthquakes might hurt the population, economy, and environment in many different ways, from destruction of buildings, lifelines, and other constructions, to triggering other natural disasters, economic and political crises. That diversity of damage requires a hierarchy of preparedness measures, from public safety legislation and insurance to simulation alarms, to preparedness at home, and red alert. Different measures can be implemented on different timescales, from seconds to decades. They should be implemented in areas of different size, from selected sites to large regions; can be maintained for different time periods; and belong to different levels of jurisdiction, from local to international. Such measures might complement, supersede, or mutually exclude each other. For this reason, optimizing preparedness involves comparison of a large number of combinations of possible measures (Davis et al., 2007, 2010, 2012).

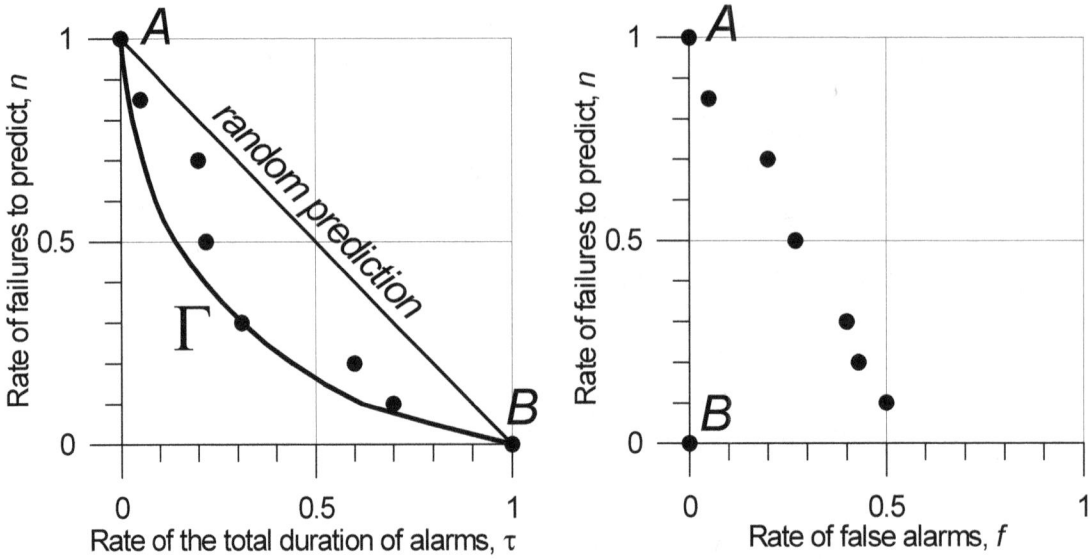

Figure 2. Error diagram. Each point shows the performance of a prediction method: the rate of failures to predict, n, the relative duration of alarms, τ, and the rate of false alarms, f. Different points correspond to different algorithms. The diagonal in the left plot corresponds to the random guess. Point A corresponds to the trivial "optimistic" strategy, when an alarm is never declared; point B marks the trivial "pessimistic" strategy, when an alarm takes place all the time; other points correspond to non-trivial predictions. Best combinations (n, τ) lie on the envelope of these points Γ. After Molchan (1997).

Disaster management has to take into account the cost/benefit ratio of possible preparedness measures. No single measure alone is sufficient. On the other hand, many efficient measures are inexpensive and do not require high accuracy of prediction. As is the case for all forms of disaster preparedness, including national defense, a prediction can be useful if its accuracy is known, even if it is not high.

Decision depends on specific circumstances in the area of alarm. At the same time, it also depends on the prediction quality, i.e. rate of failures-to-predict, n, rate of false alarms, f, and the fraction of time-space occupied by all alarms together, τ. These values are determined above. The choice of preparedness measures is never unique. Different measures may supersede or mutually exclude one another, leaving certain freedom of choice to a decision-maker (Keilis-Borok et al., 2004; Davis et al., 2007, 2010, 2012). The designer of a prediction algorithm has certain freedom to choose the trade-off between different characteristics of its accuracy (rate of failures to predict, duration of alarms, and rate of failures to predict) by varying adjustable elements of the algorithm.

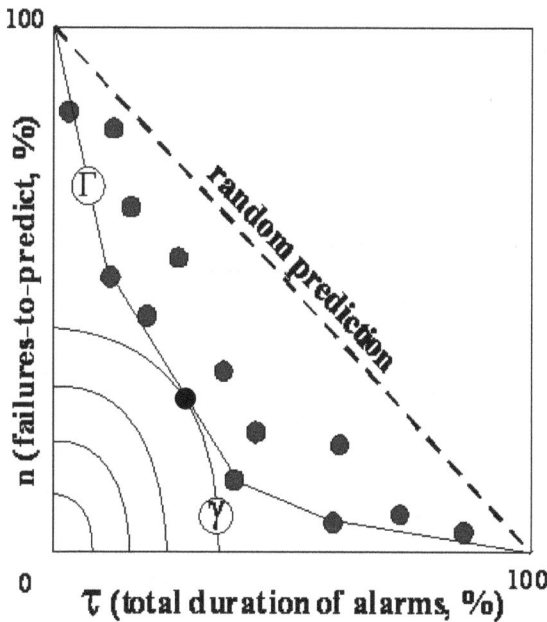

Figure 3. Joint optimization of prediction and preparedness. After Molchan (1997).

Accordingly, prediction and preparedness should be optimized jointly; *there is no "best" prediction per se* (Molchan, 1997, 2003; Molchan and Keilis-Borok, 2008). A framework for such optimization is shown in Fig. 3. Dots show points on an error diagram, Γ is their envelope. The contours show "loss curves" with constant value of prevented damage γ (see *Molchan*, 1997). Optimal strategy is the tangent point of contours Γ and γ. Thus disaster preparedness would be more flexible and efficient if prediction would be carried out in parallel with several versions of an algorithm. This has not yet been done.

Other software and know-how for assisting decision-makers in choosing the best combination is described by Kantorovich et al. (1974), Molchan (1991, 2003) and Keilis-Borok (2003). A hypothetical example is given in Fig. 4 (Davis et al., 2007). Imagine an earthquake alarm covering an area shown on the map – a part of Central California. The map shows the parts of the water supply systems in the area that are vulnerable to earthquakes. Table 2 shows the cost-efficiency of some preparedness measures (Keilis-Borok et al., 2004; Davis et al., 2007, 2010).We see that lowering the water level is justified only for a fragile reservoir and for the probability of false alarm 50% or less, but not 75%. Decision making might also require estimating distribution functions for different types of damage, casualties included (Keilis-Borok et al., 2004).

Table 2. Gain from preparedness actions for different probabilities of false alarms

	Action	DA ($1,000)	DP ($1,000)	Gain ($1,000)		
				f = 10%	f = 50%	f = 75%
T	Lower water level in Fragile Reservoir	2,000	7,500	4,750	1,750	*125*
T	Lower water level in Stout Reservoir	2,000	10	*1,991*	*1,995*	*1,998*
T	Drain Reservoirs	16,000	7,510	*9,240*	*12,250*	*14,120*

Gain G is calculated by formula $G = DP(1 - f) - DA$ where DP is damage prevented, DA – cost of action, f – probability of false alarm. T = Temporary: lasting for alarm period. Negative gain is shown in bold italic.

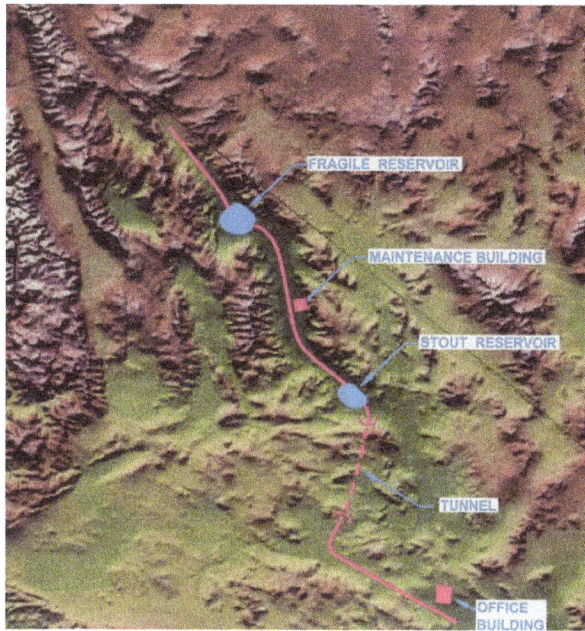

Figure 4. Schematic example: Vulnerable objects and hazards in the area of alarm.

CHAPTER 2

Earthquake Prediction

We understand earthquake prediction as prediction of time interval and geographic area where an individual future strong earthquake will occur. The strong earthquake is determined as an earthquake with magnitude $M \geq M_0$ where M_0 is a certain threshold. The magnitude range of the target earthquakes is sometimes limited by an upper boundary. Prediction is meaningful if it includes an estimated rate of false alarms. Taking into account that in the case of earthquakes the geographic area is predicted, Fig. 1 is transformed into Fig. 5.

Premonitory patterns signal destabilization of the earthquakes-prone lithosphere and thus an increase in the probability of a strong earthquake. Premonitory patterns do not necessarily contribute to causing a subsequent strong earthquake; both might be parallel manifestations of the same underlying process – the tectonic development of the Earth in multiple time-, space-, and energy- scales. For that reason premonitory patterns might emerge in a broad variety of observable fields reflecting lithosphere dynamics, and in different scales.

The process of earthquake prediction consists of consecutive, step-by-step, narrowing down of the time interval, space, and magnitude ranges where a strong earthquake has to be expected. Five stages of prediction are usually distinguished. The ***background stage*** provides maps with territorial distribution of maximum possible magnitude and recurrence time of destructive earthquakes having different magnitudes. The four subsequent stages, loosely divided, include the time-prediction; they differ in the characteristic time intervals covered by an alarm.

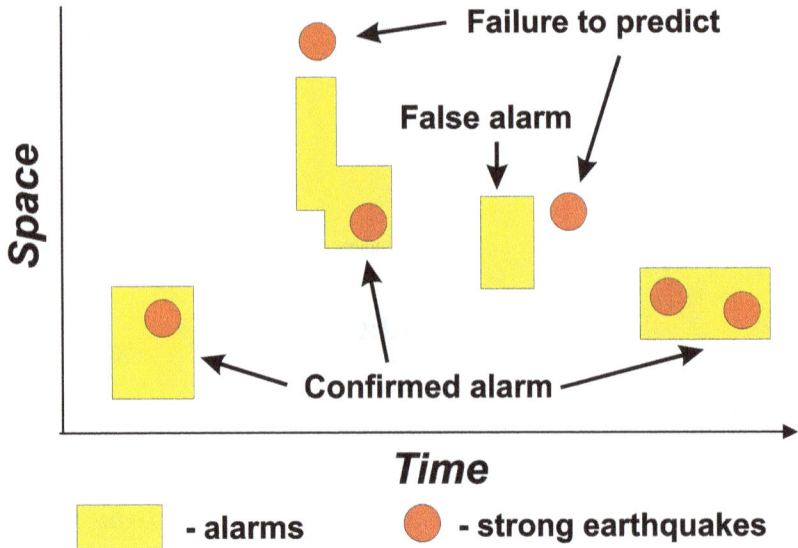

Figure 5. Possible outcomes of earthquake prediction.

These stages are as follows:
- **long-term** (10^1 years),
- **intermediate-term** (years),
- **short-term** (10^{-1} to 10^{-2} years), and
- **immediate** (10^{-3} years or less).

Such division into stages is dictated both by the characteristics of the process leading to a strong earthquake and by the needs of the earthquake preparedness in each particular region. The latter comprises an arsenal of safety measures for each stage of prediction, like in preparedness to a war.

2.1. An early example

The first premonitory seismicity pattern formally defined and featuring long-range correlations and worldwide similarity was the *"pattern Σ"*, introduced in 1964 by Keilis-Borok and Malinovskaya (1964). It comprises the premonitory increase of the total area of the ruptures in the earthquake sources in a medium magnitude range. Emergence of this pattern was captured by the function:

$\Sigma(t/s,B) = \sum 10^{Bmi}$

where m_i is the magnitude of ith earthquake; the sum $\sum$ is taken over all earthquakes that have occurred during the time interval $(t-s, t)$ within the region considered; $B \approx 1$. With this value of B, the summands are coarsely proportional to the source area (with $B = 0$ and $B = 3/2$ this sum would correspond to the number of earthquakes and their total energy, respectively).

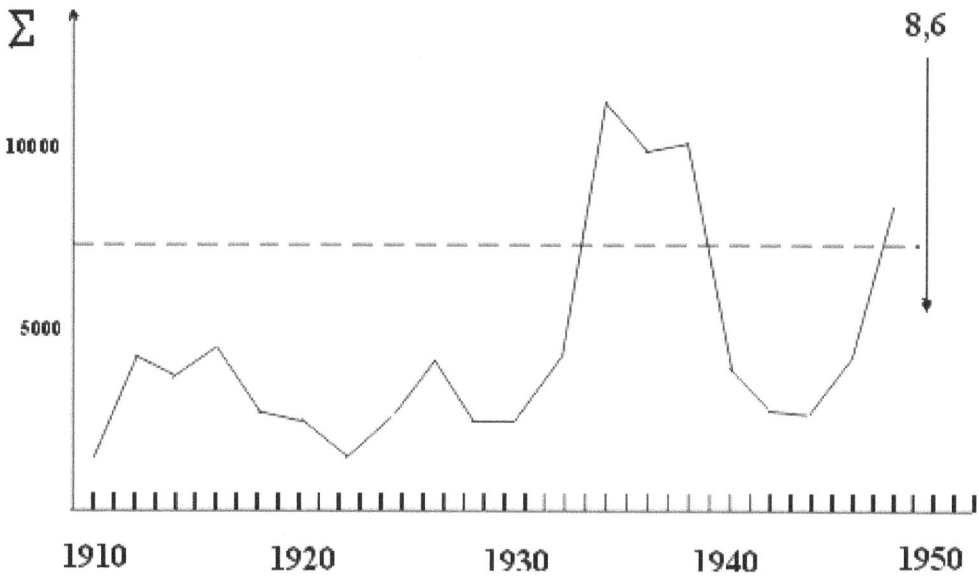

Figure 6. Illustration of the premonitory seismicity pattern Σ: rise of the function $\Sigma(t/s,B)$ before the Assam earthquake in India (1950, M = 8.6). The emergence of pattern Σ was captured by the condition $\Sigma(t) \geq C_\Sigma$, threshold C_Σ (shown by the horizontal line) was determined uniformly for all regions.

The change of the function $\Sigma(t)$ in periods preceding 20 strong earthquakes worldwide was investigated by Keilis-Borok and Malinovskaya (1964). It has been shown that the function $\Sigma(t/s,B)$ strongly increased one to ten years prior to each of the earthquakes considered, indicating *"a direct connection between strong earthquakes and the very large scale features of the development of the whole Earth's crust."* Fig. 6 shows an example for the catastrophic Assam earthquake in India, 1950, M=8.6. The emergence of pattern Σ was captured by the condition $\Sigma(t) > C_\Sigma$, with threshold C_Σ determined uniformly for all regions.

Pattern Σ was the first premonitory seismicity pattern that demonstrated the major features of patterns discovered later: long-range correlations and similarity. These features can be described as follows.

(i) *Long-range correlation*: The area of earthquake preparation can greatly exceed the source of the incipient earthquake. This is reflected in the large size of the areas that have to be used for calculation of the function Σ (Fig. 7).

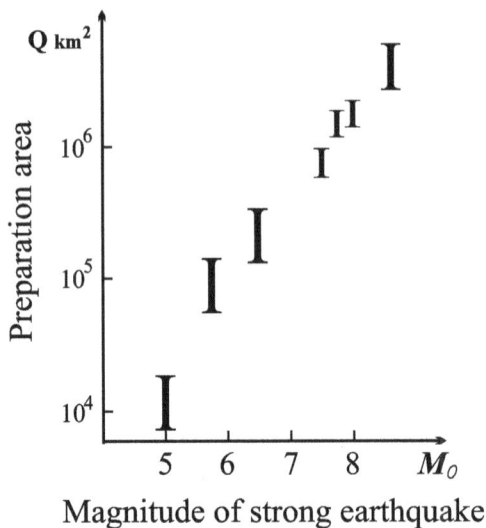

Figure 7. Long-range correlation in the formation of premonitory seismicity patterns. $Q(M_o)$ is the area, where pattern Σ was formed prior to a strong earthquake with $M \geq M_o$.

(ii) *Similarity*: Pattern Σ is self-adapting; it has a uniform definition for different magnitude thresholds M_0 of strong earthquakes targeted for prediction. Specifically,

— the area of preparation is a power-law function of M_0 (Fig. 7); and

— the threshold for identification of pattern Σ is normalized by M_0, $C_\Sigma = 0.5 \times 10^{BM0}$; this means that the area unlocked by medium magnitude earthquakes reached at least half of the area which will be unlocked by an incipient strong earthquake.

These features have been confirmed by subsequent studies and eventually evolved in the prediction paradigms.

Pattern Σ illustrates the consecutive stages in the search for prediction algorithms.

— It started with the hypothetical *premonitory phenomenon*: the rise of seismic activity, which is one of the basic characteristics of seismicity.

— That phenomenon was captured by a specific *precursor*: large area unlocked by earthquakes in a medium magnitude range. The same phenomenon is captured also by other precursors, for example by the number of earthquakes not weighted by magnitude.

— Finally, the prediction algorithm was formally defined by the function $\Sigma(t/s,B)$ and the threshold C_Σ.

2.2. Algorithms M8 and MSc ("the Mendocino Scenario")

The intermediate-term earthquake prediction algorithm M8 was designed by retroactive analysis of dynamics of seismic activity preceding the greatest, magnitude 8.0 or more, earthquakes worldwide (hence its name). Its prototype (Keilis-Borok and Kossobokov, 1984) and the original version (Keilis-Borok and Kossobokov, 1987) were tested retroactively at 143 points, of which 132 are recorded epicenters of earthquakes of magnitude 8.0 or greater from 1857-1983.

The algorithm is aimed at prediction of earthquakes with magnitude M_0 and higher from the range $M_0+ = [M_0, M_0+\Delta M]$ (where $\Delta M < 1$). Data used by the algorithm is a catalog of main shocks: $\{t_i, m_i, h_i, b_i(e)\}$, $i = 1, 2, ...$ Here t_i is the origin time, $t_i \leq t_{i+1}$; m_i is the magnitude, h_i is focal depth, and $b_i(e)$ is the number of aftershocks with magnitude M_{aft} or more during the first e days after a relevant main shock. Overlapping circles, with the diameter $D(M_0) = (exp(M_0-5.6)+1)^\circ$ in degrees of the Earth meridian (this is 384, 560, 854 and 1333 km for M_0 = 6.5, 7.0, 7.5, and 8, respectively), scan the seismic region under study. Only main shocks with $M_0 > m_i \geq M_{min}$ are considered where M_{min} is determined by the condition that the average annual number of main shocks with $m_i \geq M_{min}$ in the circle is equal to $\tilde{N}$. The main shocks are used to compute functions in the sliding time window $(t - s, t)$. The functions depict different measures of intensity in earthquake flow, its deviation from the long-term trend, and clustering of earthquakes. The functions include:

- $N(t) = N(t \mid M_{min}, s)$, the number of main shocks of magnitude M_{min} or larger in $(t - s, t)$;
- $L(t) = L(t \mid M_{min}, s, t_0)$, the deviation of $N(t)$ from the longer term trend, $L(t) = N(t) - N_{cum}(t - s)(t - t_0) / (t - t_0 - s)$, where $N_{cum}(t) = N(t \mid M_{min}, t - t_0)$ is the cumulative number of main shocks with $M \geq M_{min}$ from the beginning of the sequence t_0 to t;
- $Z(t) = Z(t \mid M_{min}, M_1, s, g, a, \beta)$, linear concentration of main shocks $\{i\}$ from the magnitude range (M_{min}, M_1), $M_1 = M_0 - g$ and interval $(t - s, t)$; the linear concentration is estimated as the ratio of the average source diameter l to the average distance r between sources, where source diameter of an earthquake with magnitude M is estimated by $10^{\beta(M - a)}$; and
- $B(t) = B(t \mid p, q, s_1, M_{aft}, e) = \max_{\{i\}}\{b_i(e)\}$, the maximum number of aftershocks (i.e., a measure of earthquake clustering); the sequence $\{i\}$ is considered in the trailing time window $(t - s_1, t)$ and in the magnitude range $(M_0 - p, M_0 - q)$.

Each of the functions N, L, and Z is calculated twice with M_{min} determined for $\tilde{N} = 10$ and $\tilde{N} = 20$. As a result, the earthquake sequence is given a robust averaged description by seven functions: N, L, Z (twice each), and B.

"Very large" values are identified for each function from the condition that they exceed Q percentiles (i.e., they exceed Q percent of the encountered values).

An alarm or a TIP, time of increased probability, is declared for 5 years when at least six out of seven functions, including B, become "very large"' within a narrow time window $(t - u, t)$. To stabilize prediction, this criterion is checked at two consecutive moments, t and $t + 0.5$ years. In the course of a forward application, the alarm can extend beyond, or terminate in less than 5 years when updating causes changes in the magnitude cutoffs and/or the percentiles.

The following standard values of parameters indicated above are prefixed in the algorithm M8: $s = 6$ years, $s_1 = 1$ year, $g = 0.5$, $p = 2$, $q = 0.2$, $u = 3$ years, and $Q = 75\%$ for B and 90% for the other six functions. Usually, the average diameter l of the source is estimated by $(n)^{-1} \sum_{\{i\}} 10^{\beta(M_i - a)}$, where n is the number of main shocks in $\{i\}$, $\beta = 0.46$ to represent the linear dimension of source, and $a = 0$ (which does not restrict generality), and the average distance r between sources is set proportional to $n^{-1/3}$. The performance of the algorithm can be improved by estimating the linear concentration of main shocks more accurately.

Running averages are defined in a robust way, so that a reasonable variation of parameters does not affect predictions. At the same time, the discrete character of seismic data and strict usage of the prefixed thresholds result in a certain discreteness of the alarms.

It is worth mentioning that, qualitatively, algorithm M8 uses a rather traditional description of a dynamic system, adding dimensionless concentration (Z) and a characteristic measure of clustering (B) to the phase space of rate (N) and rate differential (L). The algorithm recognizes a *criterion* defined by extreme values of the phase space coordinates, as the vicinity of a system singularity. When a trajectory enters the criterion, the probability of an extreme event increases to a level sufficient to provide it effectively. The choice of the M8 criterion determines a specific intermediate-term rise, an inverse cascade (Gabrielov et al, 2000a), of seismic activity at the middle-range distance.

The second approximation prediction method "the Mendocino Scenario" (MSc) was designed by retrospective analysis of the detailed regional seismic catalog prior to the Eureka earthquake (1980, $M = 7.2$) near Cape Mendocino in California (Kossobokov et al., 1990). Given a TIP diagnosed for a certain territory ***U*** at time ***T***, the algorithm is designed to find a smaller area ***V*** within ***U***, where the predicted earthquake can be expected. To execute the algorithm, one needs a reasonably complete catalog of earthquakes with magnitudes $M \geq M_0 - 4$, which is lower than the minimum threshold usually used by M8. When this condition does not hold, we assume that the dynamics of earthquakes available in the database inherits behavior from lower levels of the seismic hierarchy. The detection of the MSc criteria in such a case is more difficult and might result in additional failures to predict.

The essence of MSc can be summarized as follows. Territory ***U*** is coarse-grained into small squares of size $s \times s$. Let's identify the squares by their sequence numbers $j = 1, 2, ..., J$. Within each square j, the number of earthquakes $n_j(k)$, aftershocks included, is calculated for consecutive, short, time windows u, months long, starting from time $t_0 = T - 6$ years onward, to include earthquakes that contributed to the TIP's diagnosis; k is the sequence number of a time window.   In this way, the time-space considered is divided into small boxes (j, k) of size $(s \times s \times u)$. *"Quiet" boxes j* are defined by the condition that $n_j(k)$ is below the Q percentile of n_j. The clusters of q or more quiet boxes connected in space or in time are identified. Area **V** is the territorial projection of these clusters.

The standard values of parameters adjusted for the 1980 Eureka earthquake are as follows: $u = 2$ months, $Q = 10\%$, $q = 4$, and s = $3D/16$, where D is the diameter of the circles used in the M8 algorithm.

Table 3. Scoring of M8 and M8 & MSc predictions, 1992-2010

Algorithm	Total number of target earthquakes	Number of predicted earthquakes	Space-time volume of alarm
M8	17	12	29%
M8 & MSc	17	8	15%

Qualitatively, the MSc algorithm outlines such an area of the territory of an alarm where the activity, from the beginning of the seismic inverse cascade recognized by the M8 algorithm in declaration of the alarm, is continuously high, but is infrequently interrupted for a short time. Such interruption must have a sufficient temporal and/or spatial span. The phenomenon, which is used in the MSc algorithm, might reflect the second (possibly, shorter term and, definitely, narrow-range) stage of the premonitory rise of seismic activity near the incipient source of a main shock. The anomalous quiescence used in the definition of the *precursory intermittent pattern* in the dynamics of the seismic region should not be mixed with a prolonged state of *"seismic quiescence"*, advocated by (Wyss and Habermann, 1988).

Scoring. Thus far, the algorithms have had the most success in predicting future earthquakes in the magnitude range 8-8.5 (Table 3). Statistical significance of these predictions exceeds 99%.

2.3. Other Premonitory Phenomena and Algorithm RTP

One premonitory phenomenon depicting a change in the spatial distribution of events prior to strong earthquakes was found in Southern California and then in the Lesser Antilles and on the Dead Sea Transform. Named **"seismic reversal" (SR)**, it characterizes a period of opposite seismic activity in areas complementary to the bulk of long-term territorial distribution of low-magnitude earthquakes (Keilis-Borok et al., 1994; Shebalin et al., 1996; Shebalin and Keilis-Borok, 1999). The phenomenon happens several months prior to a strong earthquake in the area surrounding its future source. The diameter of the prediction area is of the order of 100 km.

The other two phenomena, named ROC and Accord, were designed in inspiration of evident patterns in a lattice-type "Colliding Cascades" model of interacting elements (Gabrielov et al., 2000a, 2000b). Pattern ROC describes an increase in the earthquake correlation range a few days before a strong earthquake. It was found in the Lesser Antilles (Shebalin et al., 2000), where the corresponding prediction algorithm demonstrated extreme efficiency unusual even for retrospective analysis. The pattern Accord depicts spreading of seismic activity over a fault network and is defined as a simultaneous rise of seismic activity in a sufficiently large number of neighboring fault zones. In Southern California, scaled to target magnitude 7.5, the pattern Accord emerges within a few years in the approach of each of the three largest earthquakes: Kern County, 1952; Landers, 1992; and Hector Mine, 1994, and at no other time.

The only way to answer the question whether the phenomena are really premonitory is to define prediction algorithms on their basis and to test their performance in advance prediction. All three prediction methods are fully reproducible and presented as hypotheses for testing in advance prediction. Below we describe each of them in more detail.

"Reverse Tracing of Precursors" (RTP) algorithm (Shebalin et al., 2004, 2006; Keilis-Borok et al., 2004a). The algorithm is aimed at predictions about 9 months in advance, much shorter than by the M8 algorithm. This algorithm, as its name suggests, traces precursors in the reverse order of their formation. First it identifies "candidates" for short-term precursors. These are long, quickly formed chains of earthquakes in the background seismicity. Such chains reflect an increase in the earthquake correlation range (Fig. 8). Next, each chain is examined to determine whether there had been any preceding intermediate-term precursors in its vicinity within the previous five years. If so, the chain triggers an alarm.

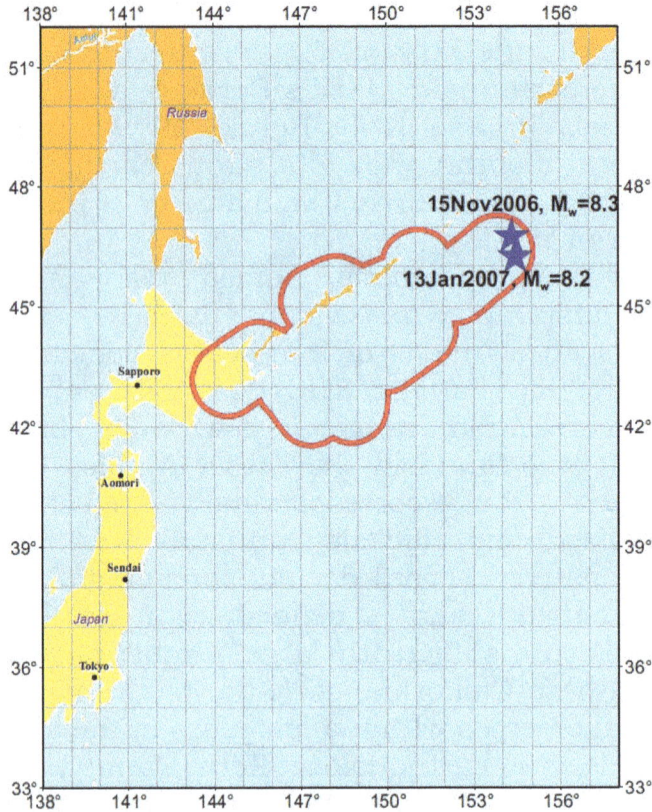

Figure 8. RTP Prediction of the 2006-2007 Simushir earthquakes in Kuril Islands. Red contour – area of alarm issued on October 2006 for 9 months. This was confirmed by the earthquake of November 15, 2006, M=8.3, followed by the earthquake of January 13, 2007, M=8.2 (the sites are denoted by stars).

The RTP algorithm has been tested by predicting future earthquakes in five regions of the world (California and adjacent regions; Central and Northern Italy with adjacent regions; Eastern Mediterranean; Northern Pacific, Japan and adjacent regions). Out of 19 alarms, five were correct and fourteen false, two of the latter being near misses occurring close to alarm areas. Five out of seven target earthquakes have been predicted (captured by alarms) and two missed. The data are still insufficient for rigorous estimation of statistical significance.

2.4 Four Paradigms in Earthquake Prediction

In the wake of the developments in the earthquake prediction, the following four paradigms have been established at the crossroad between exploratory data analysis, statistical physics, and the dynamics of fault networks (Keilis-Borok, 1994, 1996).

I. *Basic types of premonitory phenomena* comprising the variation in relevant observable fields.

II. *Long-range correlations in fault system dynamics.* Premonitory phenomena are formed not only in the vicinity of the incipient source, but also within a much wider area.

III. *Partial similarity of premonitory phenomena* in the diverse conditions, from fracturing in laboratory samples to major earthquakes worldwide, and possibly even to starquakes.

IV. *The dual nature of premonitory phenomena.* Some of them are "universal," common for complex non-linear systems of different origin; others are Earth-specific.

The paradigms discussed here have been found in the quest for premonitory seismicity patterns in the observed and modeled seismicity. There are compelling reasons to apply them also to premonitory phenomena in other relevant fields.

First Paradigm: Basic types of premonitory phenomena.
The approach of a strong earthquake is indicated by the following changes in the basic characteristics of seismicity:

(i) Rise of seismic activity;

(ii) Rise of earthquake clustering in space and time;

(iii) Rise of the earthquake correlation range;

(iv) Transformation of magnitude distribution (Fig. 9);

(v) Rise of irregularity in space and time;

(vi) Reversal of territorial distribution of seismicity;

(vii) Rise of correlation between different components (decrease of dimensionality);

(viii) Rise of response to excitation.

Other relevant processes exhibit premonitory phenomena of the same types.

Patterns of the first two types, (i) and (ii), were found first in observations (Gabrielov et al., 1986; Keilis-Borok, 1990; Keilis-Borok and Shebalin, 1999) and then in models (e.g., Gabrielov et al., 2000a; Shnirman and Blanter, 2003; Soloviev and Ismail-Zadeh, 2003); patterns of the next three types, (iii) – (v), were found in the reverse order, first in models (Gabrielov et al., 2000b; Newman et al., 1995; Sornette, 2000; Shnirman and Blanter, 2003) and then in observations (Sammis et al., 1996; Kossobokov and Shebalin, 2003); reversal of territorial distribution of seismicity (vi) was found in observations and not explored yet through modeling (Kossobokov and Shebalin, 2003); the last two phenomena remain purely hypothetical so far.

Validation. Patterns of the first two types, rise of intensity and clustering, have been validated by statistically significant predictions of real earthquakes (Molchan et al., 1990; Kossobokov and Shebalin, 2003); other patterns are undergoing different stages of testing.

Reminiscence of theoretical physics. The premonitory phenomena listed above bear a resemblance to the asymptotic behavior of a nonlinear system near the point of phase transition of the second kind. However, the problem under consideration is unusual for statistical physics: the state considered is not the equilibrium state, but the *growing disequilibrium* culminated by a critical transition.

Seismicity patterns. Premonitory phenomena of each type are depicted by different seismicity patterns. Intermediate-term patterns, with characteristic duration of alarms equaling years, are the ones systematically explored. A few examples follow.

— *Measures of seismic activity*: total area of ruptures in the earthquake sources (Sect. 2.1); accumulated strain release (Bufe and Varnes, 1993; Varnes, 1989; Bowman et al., 1998); the number of earthquakes in a certain magnitude range (Wyss and Habermann, 1988; Knopoff et al., 1996; Kossobokov and Shebalin, 2003); the time period when a given number of earthquakes occurs (lower time obviously indicates higher activity) (Sykes and Jaumé, 1990), etc.

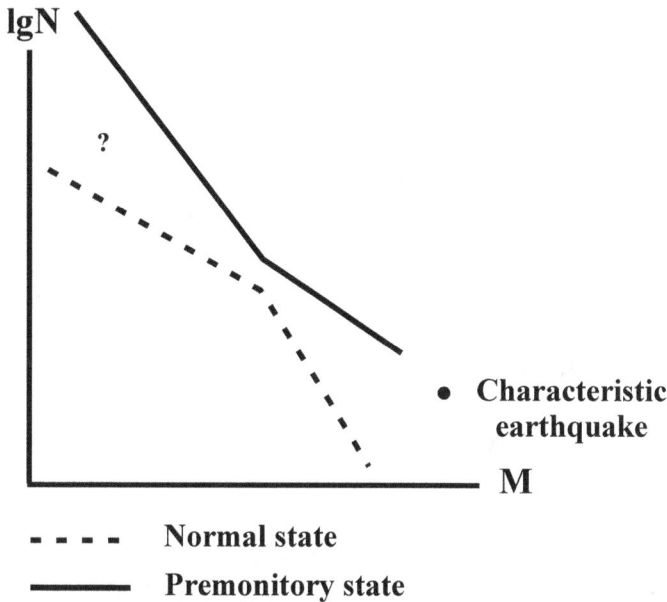

Figure 9. Scheme of premonitory transformation of the Gutenberg–Richter relation $N(M)$. N is the number of earthquakes of magnitude M or larger. Dashed and solid lines correspond to time intervals far from a strong earthquake and close to it, respectively.

— *Measures of earthquake clustering*: the number of aftershocks closely following a medium magnitude main shock ("bursts of aftershocks," see Molchan et al., 1990; Kossobokov and Shebalin, 2003); swarms of the main shocks of medium magnitude (Caputo et al., 1977; Keilis-Borok et al., 1980; Keilis-Borok et al., 1982); swarms of relatively small earthquakes (Caputo et al., 1983) etc.
— *Measures of earthquake correlation range*: the distance between nearly simultaneous earthquakes and the number of faults with nearly simultaneous rise of activity (patterns "*ROC*" and "*Accord*", Shebalin et al., 2000; Keilis-Borok et al., 2002; Zaliapin et al., 2002); the distribution of link lengths in a single

link cluster connecting earthquakes occurred in a given spatiotemporal region (Zoller et al., 2001).

— *The measures of irregularity*: the variation in magnitudes or strain release (Newman et al., 1995; Sornette and Sammis, 1995; Sornette, 2000).

— *The measures of premonitory change of magnitude distribution* (Gutenberg-Richter relation). That transformation is schematically illustrated in Fig. 9. One of its measures is the slope of the magnitude distribution in a relatively high magnitude range (pattern "Upward Bend;" Shnirman and Blanter, 2003); another measure is the difference between its slopes for lower and higher magnitudes ("pattern γ," Rotwain et al., 1997). The question mark in the figure indicates that "pattern γ," the reversal of curvature of the distribution, is yet less tested than the "Upward Bend."

Why different measures are used for the same premonitory phenomenon? These measures are certainly correlated, even by definition. Several of them are used instead of an "optimal" one for the following reasons.

— A premonitory phenomenon may have different manifestations on different timescales, spatial scales, and magnitude ranges.

— A set of measures is more reliable than a single one, due to the complexity of the processes considered, and to unavoidable noise.

— In lieu of an adequate theory, premonitory seismicity patterns have been found by heuristic analysis, and more compact definitions might just be overlooked so far.

Second Paradigm: Long-range correlations. The generation of an earthquake is not localized around its future source. A flow of earthquakes is generated by a fault network, rather than each earthquake being generated by a segment of a single fault. Accordingly, the signals of an approaching earthquake come not from a narrow vicinity of the source, but from a much wider area.

Size of areas where premonitory phenomena are formed. Let M and $L(M)$ be the earthquake magnitude and the characteristic length of its source, respectively. In the intermediate-term stage of prediction (on a timescale of years) that size may reach $10L(M)$; it might be reduced to $3L(M) - L(M)$ in a second approximation (Kossobokov and Shebalin, 2003). In the long-term stage, on the timescale of tens of years, that size reaches about $100L(M)$. For example, according to Press and Allen (1995) the Parkfield (California) earthquake, with M about 6 and $L(M) \approx 10$ km "... *is not likely to occur until activity picks up in the Great Basin or the Gulf of California, about 800 km away.*"

Historical perspective. An early, and probably the first estimation of the area where premonitory patterns are formed, was obtained for pattern Σ (Fig. 7). It is noteworthy that Charles Richter, who was generally skeptical about the feasibility of earthquake prediction, made an exception for that pattern, specifically because it was based on long-range correlations. He wrote (Richter, 1964): "*... It is important that (the authors) confirm the necessity of considering a very extensive region including the center of the approaching event. It is very rarely true that the major event is preceded by increasing activity in its immediate vicinity.*" Table 4 shows similar estimations for other intermediate-term patterns.

At the same time, long-range correlations have been often regarded as counterintuitive in earthquake prediction research on the ground that redistribution of stress and strain after an earthquake in simple elastic models would be confined to the vicinity of its source ("Saint Venant principle"). Sometimes that prompted the objection: "*the earthquakes cannot trigger each other at such distances.*" The answer is that the earthquakes involved in long-range correlation do not trigger each other, but reflect the underlying large-scale dynamics of the lithosphere. More specific explanations follow.

Table 4. Estimations of the size of the area where premonitory seismicity patterns emerge

Measure	Year	$Q(L)^b$	Reference
Area of fault breaks	1964	~ 10L	Keilis-Borok and Malinovskaya (1964), Kossobokov and Shebalin (2003)
Distant aftershocks	1975	~ 10L	Prozorov (1975)
Earthquake swarms	1977	~ 5L -10L	Caputo et al. (1977)

Bursts of aftershocks, area of fault breaks, swarms	1980	~ 5L -10L	Keilis-Borok et al. (1980), Kossobokov and Shebalin (2003)
Algorithm CN[a]	1983	~ 5L -10L	Keilis-Borok and Rotwain (1990), Kossobokov and Shebalin (2003)
Algorithm M8[a]	1985	~ 5L -10L	Keilis-Borok and Kossobokov (1990), Kossobokov and Shebalin (2003)
Benioff strain release	1989	~ 5L	Varnes (1989), Bowman et al. (1998), Jaumé and Sykes (1999)
Algorithm SSE	1992	~ 5L	Levshina and Vorobieva (1992)
Number of earthquakes	1995	~ 100L	Press and Allen (1995)
Number of earthquakes	1996	~ 5L	Knopoff et al. (1996)
Near-simultaneous pairs of earthquakes	2000	~ 3L	Shebalin et al. (2000)
Correlation length via single link cluster	2001	~ 5L	Zoller et al. (2001)
Simultaneous activation of fault branches	2002	~ 10L	Zaliapin et al. (2002)

[a] Reference is given not to the original publication, but to the latest comprehensive reviews.

[b] L is the linear dimension of the approaching strong earthquake. Premonitory phenomena have been observed on a timescale of years with one exception, tens of years in the 1995 entry.

Mechanisms of long-range correlations. Correlation in earthquake occurrence at long distances greatly exceeding earthquake source dimensions is the prominent features of seismicity dynamics; it is not confined to earthquakes precursors. Among the manifestations of that correlation are the following phenomena: simultaneous changes of seismic activity within large regions (Romanowicz, 1993), migration of earthquakes along fault zones (Mogi, 1968; Vil'kovich and Shnirman, 1983; Ma et al., 1990; Kuznetsov and Keilis-Borok, 1997), and alternate rise of seismicity in distant areas (Press and Allen, 1995) and even in distant tectonic plates (Romanowicz, 1993). Global correlations have also been found between major earthquakes and other geophysical phenomena, such as Chandler wobble, variations of magnetic field, and the velocity of Earth's rotation (Press and Briggs, 1975; Keilis-Borok and Press, 1980). Several mechanisms (not mutually exclusive) have been suggested to explain the long-range correlations. They may be divided into two groups:

(i) Some explanations attribute long-range correlations to a large-scale process controlling stress and strength in the lithosphere. Such mechanisms act under different circumstances, separately or jointly. Being rather common, they make long-range correlations inevitable. Among such processes are the following:

• Microrotation of tectonic plates (Press and Allen, 1995) and crustal blocks (Soloviev and Ismail-Zadeh, 2003; Vorobieva and Soloviev, 2005); microfluctuations in the direction of mantle currents (Soloviev and Ismail-Zadeh, 2003). Each of them creates redistribution of normal and tangential stress and, consequently, redistribution of strength through a large part of the fault network.
• Migration of pore fluids in fault systems (Barenblatt et al., 1983) affects the lithosphere strength in the following ways: lubrication; stress corrosion and destabilization waves; and redistribution of hydrostatic pressure between the solid and fluid components of the fault zone.
• Hydrodynamic waves in the upper mantle (Pollitz et al., 1998) that propagate through thousands of kilometers during the decades and may trigger strong earthquakes connecting seismicity across the globe.
• Activity of creep fractures in the ductile part of the lithosphere. Deformation in the ductile part increases the

stress in the brittle part thus triggering earthquakes (Aki, 1996).

- Inelasticity and inhomogeneity of the lithosphere (Barenblatt, 1996). Due to either of the mechanisms, the redistribution of stress after fracture extends to much greater distances than in a homogeneous elastic media.

(ii) In another approach, the lithosphere is regarded as a non-linear chaotic system; then the long-range correlations are again inevitable, as a general feature of such systems in a near-critical state (Sornette and Sammis, 1995; Bowman et al., 1998; Turcotte et al., 2000).

Third Paradigm: Similarity. Premonitory phenomena are similar (identical after normalization) in extremely diverse environments and in a broad energy range. The similarity is not unlimited however and regional variations of premonitory phenomena do emerge.

Normalization. Earthquake sequences used in a prediction algorithm are normalized to ensure that the algorithm is self-adapting, i.e., it can be applied without re-adaptation in regions with different seismic regimes.

The area to which the sequence belongs to is normalized by $L(M)$ as described in Table 1.

The magnitude range is re-adapted by changing the minimum magnitude considered (e.g., Keilis-Borok and Kossobokov, 1990; Keilis-Borok and Rotwain, 1990; Kossobokov and Shebalin, 2003).

The time scale in most prediction algorithms does not depend on M; according to the Gutenberg–Richter relation, earthquakes of smaller magnitudes occur more frequently. This is not a contradiction, because the Gutenberg–Richter relation refers to a given region — the same for all magnitudes — whereas prediction is made for an area of a size proportional to $L(M)$.

This difference has a consequence which is not always recognized. Let $T_r(M)$ and $T_a(M)$ be the average return times of an earthquake of magnitude M in the whole region, and in a smaller area of linear size $L(M)$, respectively. According to the well known relations $T_r(M) \sim 10^{bM}$, and $L(M) \sim 10^{cM}$, and $T_a(M) \sim 10^{(b-cv)M}$. Here b is the slope of the Gutenberg–Richter relation, c determines the connection between the magnitude and the source dimension, and v is the fractal dimension of the cloud of epicenters. The existing estimations of parameters b (about 1), c (between 0.5 and 1), and v (between 1.2 and 2) do not contradict the hypothesis that the expression $(b - cv)$ is close to 0; accordingly, *earthquakes of different magnitudes might have about the same recurrence time in their own cells.*

Applications. Prediction algorithms, thus normalized, retain their predictive power in many cases: microfractures in laboratory samples; induced seismicity; earthquakes in subduction zones, major strike-slip fault zones, rift zones and platforms (Gorshkov et al., 1997; Keilis-Borok and Shebalin, 1999; Kossobokov and Shebalin, 2003). The corresponding seismic energy release ranges from a few ergs to 10^{25} ergs. However, performance of prediction algorithms does vary from case to case.

Frontiers of similarity, the neutron star (Kossobokov et al., 2000a). An opportunity to explore, albeit qualitatively, the frontiers of similarity is provided by registration of the 111 flashes of energy radiated from the neutron star with celestial coordinates 1806-20 in the frequency band of soft ☐-rays. These flashes have been probably originated by "starquakes," i.e. fractures in the crust of the neutron star. Environments of starquakes and earthquakes are summarized in Table 5.

Table 5. Environments of starquakes and earthquakes

Characteristics	Neutron star 1806-20	Earth
Composition of the crust	Lattice of heavy nuclei	Grains of rock
Radius	10 km	6371 km
Thickness of the crust	1 km	About 33 km
Density	10^{14} g/m³	$5.5 \square 10$ g/m³
Energy release	Up to 10^{46} erg	Up to 10^{25} erg
Driving forces	Magnetic field	Convection

Fig. 10 compares the emergence of premonitory patterns before a major starquake and the strong earthquake in the Aquaba gulf, 11 Dec. 1995, $M = 7.3$. The functions, capturing these patterns, are taken from the earthquake prediction algorithms (Kossobokov and Shebalin, 2003): Functions Σ, N, and Z capture the intensity of earthquake flow, *L* captures its deviation from the long-term trend, B is the measure of earthquakes clustering. The patterns work for Earth and the star, their fantastic difference notwithstanding. Only one parameter, the timescale for starquakes, had to be readjusted a posteriori.

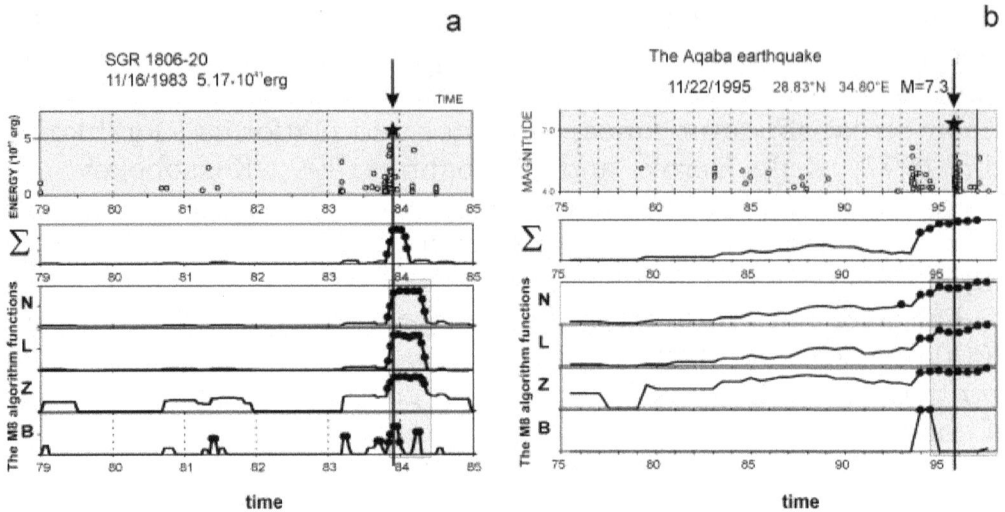

Figure 10 **Frontier of similarity of premonitory seismicity patterns.** (a) Starquake registered on 16 Nov. 1983, a coarse equivalent of its magnitude is about 20. (b) Aqaba earthquake, 22 Nov. 1995, *M* = 7.3. The panels show the sequence of starquakes (*left*) and earthquakes (*right*). *Stars* indicate major events that are targeted for prediction. Other panels show the functions capturing premonitory seismicity patterns defined in (Keilis-Borok and Kossobokov, 1990; Keilis-Borok and Rotwain, 1990; Kossobokov and Shebalin, 2003). Dots correspond to the functionals, depicting different characteristics of seismicity: *B* is the measure of clustering; other functions are different measures of seismic activity. Dots indicate the emergence of a premonitory pattern. After Kossobokov et al. (2000a).

Limitations. The performance of some prediction algorithms still does vary from region to region (see Ghill, 1994; Keilis-Borok and Shebalin, 1999; Molchan, 2003). It is not yet clear whether this is due to imperfect normalization or to Earth-specific limitations on similarity itself.

Fourth Paradigm: Dual nature of premonitory phenomena.

Some premonitory phenomena are "universal," common for hierarchical complex nonlinear systems of different origins; others are specific to the geometry of the fault networks or to a certain physical mechanism controlling the strength-stress field in the lithosphere.

A. "Universal" premonitory phenomena.

The meaning of "universality." The known premonitory seismicity patterns prove to be universal in the sense that one can reproduce them in the lattice models specific not only to Earth. Obviously, seismicity and other fields relevant to earthquake prediction are Earth specific. At the same time, these fields are subject to more general laws of nature. If it is correct, then under the same clause the premonitory phenomena connected with strong earthquakes would reflect the general laws of self-organization of complex systems. The lattice models are developed in statistical physics and nonlinear dynamics (e.g., Shnirman and Blanter, 2003).

An illustration is given in Figs. 10 – 12 (after Zaliapin et al., 2003b). The model reproduces the first four types of premonitory patterns listed above in the first paradigm. So far, this is the widest set of patterns reproduced in a model. This was done with a model of "colliding cascades" introduced by Gabrielov et al. (2000) and Zaliapin et al. (2003b).

The model describes the generation of critical phenomena by colliding cascades of loading and failures. Its major features are the following:

— The model has a ternary hierarchical structure (Fig. 11a).

— The load is applied at the top of the hierarchy and is transferred downward as a direct cascade.

— Failures are initiated at the lowest level of hierarchy and propagate upward as an inverse cascade.

— The cascades are interacting (Fig. 11b). Loading triggers failures and the failures redistribute and release the load.

Eventually the failed elements "heal," thus ensuring permanent functioning of the system.

(a)   **(b)**

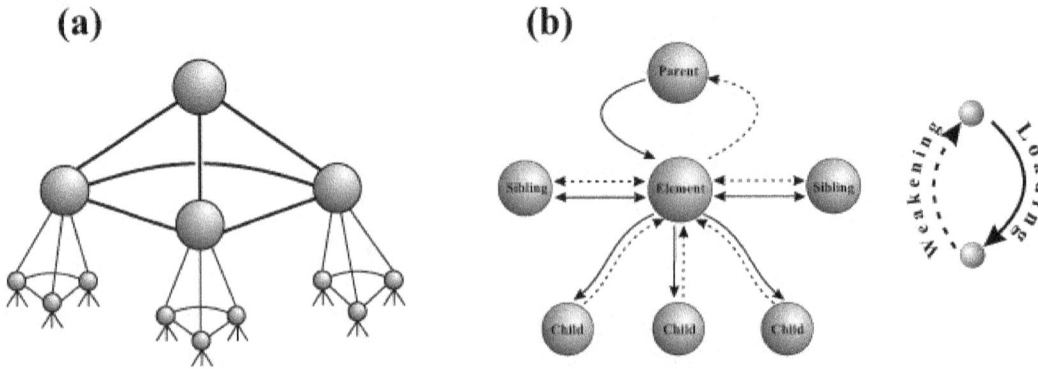

Figure 11. The structure of the colliding cascade model with branching number 3. (a) Three highest levels of the hierarchy, (b) interaction with the nearest neighbors.

In application to seismicity, the hierarchy imitates the structure of the lithosphere, where loading imitates the impact of tectonic forces, and failures imitate earthquakes; a critical phenomenon, "a major earthquake," is a failure of the top element.

Prediction of modeled seismicity. An example of an earthquake sequence generated by a colliding cascade model is shown in Fig. 12. Note that the model generates earthquakes of discrete magnitude. Prediction is targeted at the strongest synthetic earthquakes possible in the model.

Heuristic constraints. The modeled seismicity fits basic heuristic constraints, exhibiting, albeit coarsely, the major features of observed seismicity: the seismic cycle; intermittence of the seismic regime; the Gutenberg–Richter relation; aftershocks; long-range correlation; and the premonitory seismicity patterns indicated below.

Premonitory seismicity patterns. The model also reproduces the observed premonitory seismicity patterns of the first four types listed above:

(i)   rise of seismic activity;
(ii) rise of clustering;
(iii) rise of correlation range;
(iv) transformation of the Gutenberg–Richter relation.

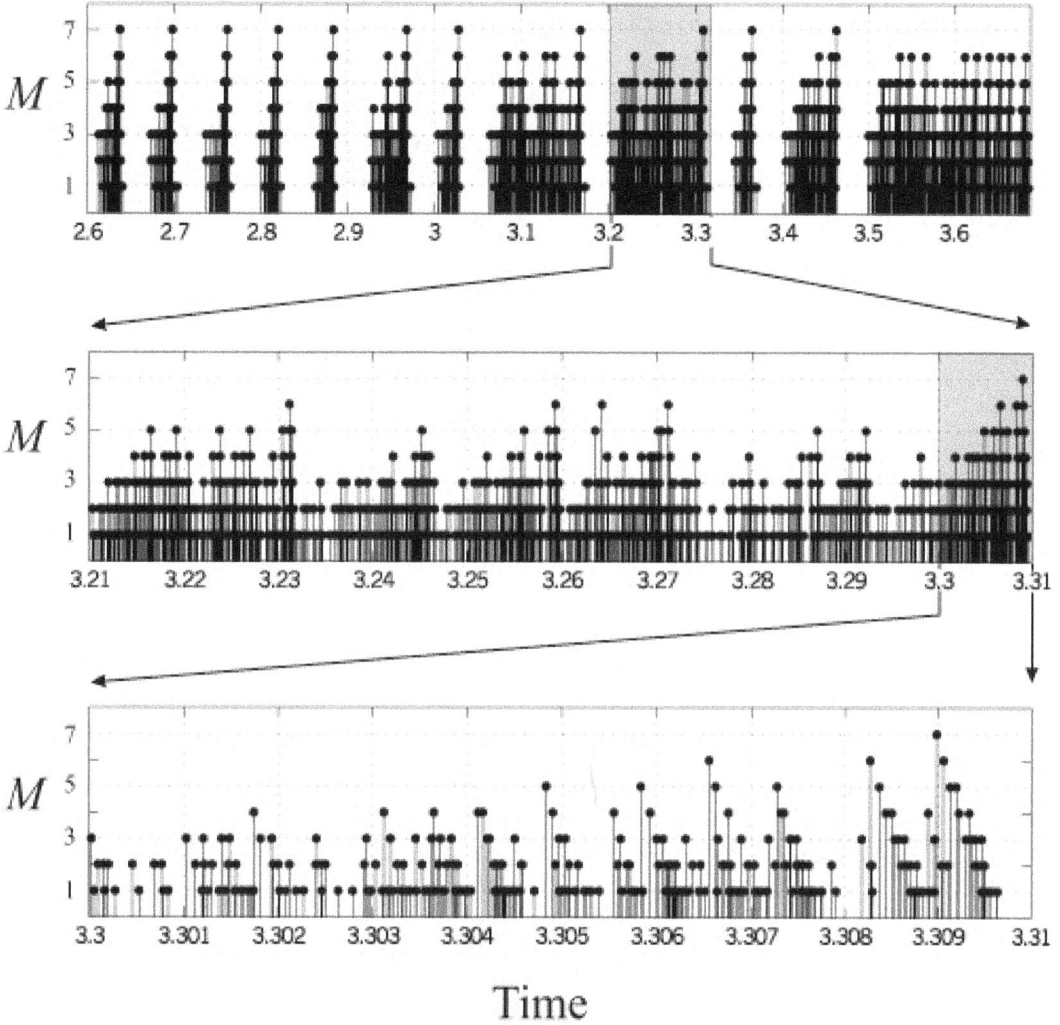

Figure 12 Synthetic earthquake sequence consecutively zoomed. Shaded areas mark zoomed intervals. The model shows the rich variety of behavior on different timescales. Note that the ratio of time scales for the top and bottom panels is 10^2.

Moreover, the premonitory rise of correlation range was found first in the colliding cascade model (Gabrielov et al., 2000a, 2000b) and then in the observed seismicity of the Lesser Antilles (Shebalin et al., 2000) and Southern California (Zaliapin et al., 2002). So far, other previously suggested patterns have not been explored with that model.

Performance. Fig. 13a shows the emergence of premonitory patterns before major earthquakes (M = 7) in the modeled seismicity. For brevity, only 10 such earthquakes are shown. Patterns of each type are defined in several magnitude ranges, so that altogether 23 patterns have been considered. Fig. 13b shows a continuous time period that includes not only major earthquakes but also long time intervals between them; here, one can see the false alarms.

The performance of these patterns in predictions of major earthquakes is estimated by Zaliapin et al. (2003b) by means of the error diagrams. In the analysis of observations this performance would be quite satisfactory. As in the analysis of observations, the collective performance of premonitory patterns is better than an individual one.

The universality of patterns is demonstrated by the fact that the model has nothing specific to only generation of earthquakes, and it can be equally interpreted in more general terms, or, for example, in terms of economic or engineering systems.

Many premonitory seismicity patterns have also been reproduced with other models of interacting elements (e.g., Pepke et al., 1994; Kossobokov and Carlson, 1995; Newman et al., 1995; Huang et al., 1998; Turcotte, 1999; Shnirman and Blanter, 2003) as well as with Earth specific models (e.g., Yamashita and Knopoff L, 1992; Panza et al., 1997; Soloviev and Ismail-Zadeh, 2003).

B. Earth-specific premonitory phenomena.

Phenomena of this kind are not yet defined in a way clear-cut enough for them to be incorporated directly in prediction algorithms. At present, we can only broadly outline the available evidence pointing at what such phenomena might be:

Geometric incompatibility G. The nodes are of obvious interest for prediction, due to their role in the nucleation of strong earthquakes and in the development of the instability of a fault network. Generally speaking, the approach of a strong earthquake is reflected in the strength-stress field, which directly controls seismicity.

(a)

8 000 time units prior to each event

(b)

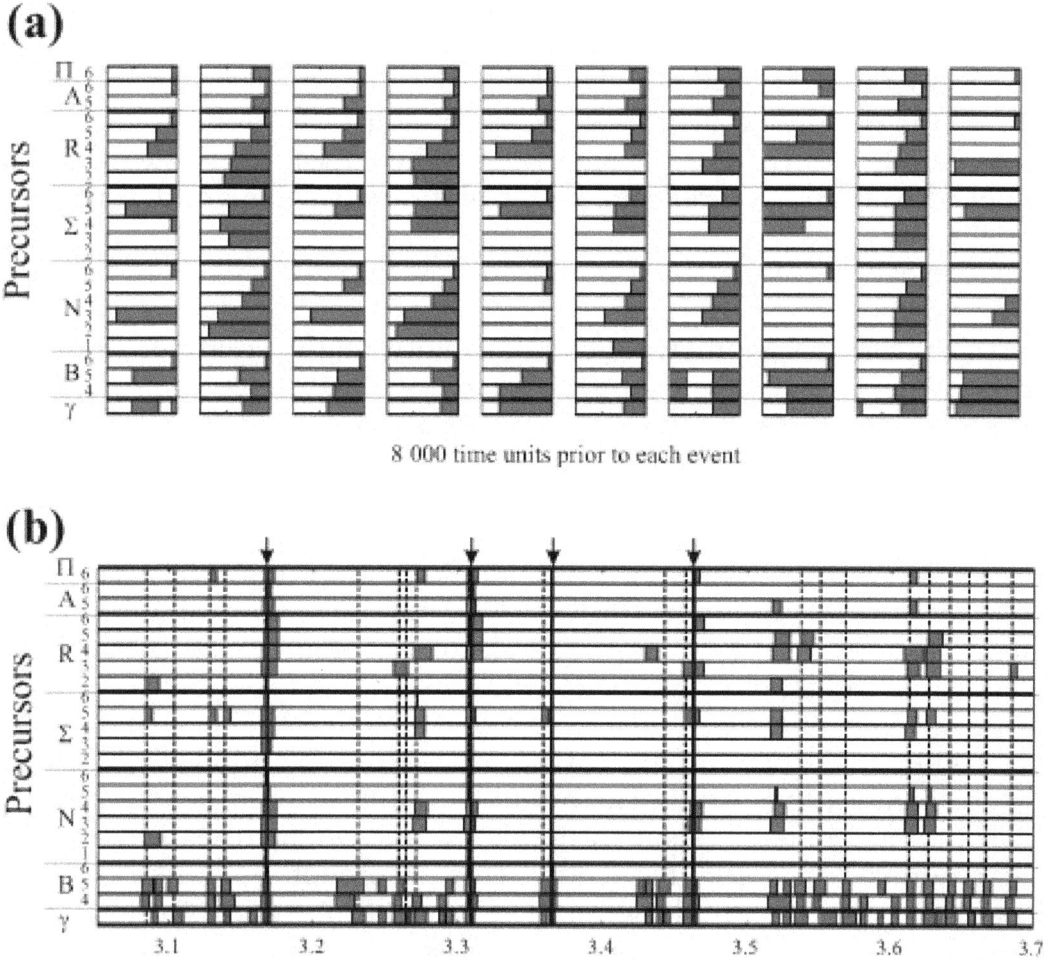

Figure 13. Collective performance of premonitory seismicity patterns (PSPs) found in observed seismicity and reproduced by colliding cascade model. After (Zaliapin et al., 2003b). The figure shows alarms generated by PSPs of the four types, discussed above: the rise of earthquake correlation range (Π_m, A_m, R_m); the rise of clustering (B_m); rise of seismic activity (Σ_m, N_m); and the transformation of the Gutenberg–Richter relation (γ). Each pattern is defined in different magnitude ranges indicated at the left side of the panel. *Shaded areas* show alarms triggered by the patterns. (a) Alarms preceding 10 major synthetic earthquakes ($M = 7$). Each box corresponds to a single major earthquake; the right edge of the box is its occurrence time. (b) Alarms during the continuous time interval. *Arrows and solid vertical lines* show the times of major earthquakes. Dashed vertical lines show the times of main shocks of magnitude $M = 6$ to illustrate their association with false alarms.

However, is hardly realistic to determine this field within a node, given its complexity. Monitoring the geometric incompatibility *G* seems to be highly promising, because:

– It is the only known control parameter that captures the state of a node relevant to seismicity; it shows whether the node is locked or unlocked; reflects accumulation of stress and fracturing, etc.

– It can be realistically monitored by observations from outside the nodes.

Reversals of geometric incompatibility might create or dissolve phenomena highly relevant to prediction (Gabrielov et al., 1996): asperities; relaxation barriers; weakest links; and alternation of seismicity and creep ("loud" and "silent" earthquakes).

If this inference is correct, these features would migrate from node to node at the large velocity typical of seismicity migration: tens to hundreds km/year (Vil'kovich and Shnirman, 1983).

Kinematic incompatibility *K*. This is a control parameter complementary to *G*. It is relevant to predictions because it captures the state of faults and the internal state of blocks within a given territory.

Where to expect a next earthquake? Both *K* and *G* may change after consecutive strong earthquakes. This opens a new approach to identification of "soon-to-break" faults, where the next strong earthquake is going to occur. This approach is illustrated by estimates of *G* in the large node in the San Andreas fault zone at the southeastern corner of the famous Big Bend (Gabrielov et al., 1996). This node lies on the junction of three groups of faults: San Andreas proper, right-lateral strike-slip faults to the north, and the thrust faults to the west. They will be called *SA*, *N*, and *W*, respectively. The impact of earthquakes in each system might be summed up as follows.

–   Earthquakes in *SA* increase the kinematic incompatibility of the node, thus continuing stress accumulation there.

–   Earthquakes in *N* or *W* act in the opposite way: they reduce the kinematic incompatibility, thus releasing stress.

This might explain why the last three major earthquakes occurred not at San Andreas fault, where it would seem natural to expect them, but in both of the adjacent fault groups, the Landers (1992, *M* = 7.6) and the Hector Mine (1999, *M* = 7.3) earthquakes on the *N*, and the Northridge earthquake (1994, *M* = 6.7) on *W*.

The very possibility of such a strong earthquake on the fault where it actually occurred was a surprise. It is noteworthy, therefore, that the epicenters of these three earthquakes lie within the nodes where, according to (Gelfand et al., 1976), earthquakes of magnitude 6.5 and above can nucleate.

Specific precursors in nodes. A much simpler but also promising possibility is to separately consider premonitory phenomena inside and outside of nodes. The difference between these two cases was observed by Rundkvist and Rotwain (1996) who applied the earthquake prediction algorithm CN in the Red Sea Rift zone. They found that during a correct alarm, the node was seismically silent, whereas during false alarms, it was active.

CHAPTER 3

Prediction of the Outcomes of the U.S. Elections

In terms of predictions, each election event, like an earthquake, can be considered as an extreme event in a socioeconomic system, with the specific that determine predictions of the election outcomes.

The elections' time is set by the law as follows.

— *National elections* are held every even-numbered year, on the first Tuesday after the first Monday in November (i.e., between November 2 and November 8, inclusively).

— *Presidential elections* are held once every 4 years, i.e. on every other election day. People in each of the 50 states and the District of Columbia are voting separately for "electors", pledged to one or another of the Presidential candidates. These electors make up the "Electoral College" which directly elects the President. Since 1860, when the present two-party system vas basically established, Electoral College reversed the decision of the popular vote only four times, in 1888, 1912, 2000, and 2016. Algorithmic prediction of such reversals is not developed so far.

— A third of *Senators* are elected for a 6-year term every election day; "mid-term' elections held in the middle of a Presidential term are considered here.

3.1. Methodology

The prediction target is an electoral defeat of an "incumbent" party, i.e. the party holding the contested seat. Accordingly, the prediction problem is formulated as whether the incumbent party will retain this seat or lose it to the challenging party (*and not whether Republican or Democrat will win*). As is shown below, this formulation is crucial for predicting the outcomes of elections considered.

Data. The pre-election situation is described by robust common sense parameters defined at the lowest (binary) level of resolution, as the *yes* or *no* answers to the questionnaires given below (Tables 6, 7). The questions are formulated in such a way that the answer *no* favors the victory of the challenging party. According to the Hamming distance analysis, the victory of the challenging party is predicted when the number D of answers *no* exceeds a threshold D^*.

Table 6. Questionnaire for mid-term Senatorial Elections (Lichtman and Keilis-Borok, 1989)

(Incumbency):   The incumbent-party candidate is the sitting senator.

(Stature):   The incumbent-party candidate is a major national figure.

(Contest):   There was no serious contest for the incumbent-party nomination.

(Party mandate):   The incumbent party won the seat with 60% or more of the vote in the previous election.

(Support):   The incumbent-party candidate outspends the challenger by 10% or more.

(Obscurity):   The challenging-party candidate is not a major national figure or a past or present governor or member of Congress.

(Opposition):   The incumbent party is not the party of the President.

(Contest):   There is no serious contest for the challenging-party nomination (the nominee gains a majority of the votes cast in the first primary and beats the second-place finisher at least two to one).

3.2. Mid-term Senatorial Elections

The prediction algorithm was developed by a retrospective analysis of the data on three elections, 1974, 1978, and 1982. The questionnaire is shown in Table 6. Victory of the challenger is predicted if the number of answers *no* is 5 or more (Lichtman and Keilis-Borok, 1989).

The meaning of these questions may be broader than their literal interpretation. For example, financial contributions (# 5) not only provide the resources required for an effective campaign, but may also constitute a poll in which the preferences are weighed by the money attached.

Predicting future elections. This algorithm (without any changes from year to year and from state to state) was applied in advance to the five subsequent elections, 1986 – 2002. Predictions are shown in Fig. 14. Altogether, 150 seats were put up for election. For each seat a separate prediction was made, 128 predictions were correct, and 22 – wrong.

Statistical significance of this score is 99.9%. In other words the probability to get such a score by chance is below 0.1% (Lichtman and Keilis-Borok, 1989; Lichtman, 1996, 2000). For some elections these predictions might be considered as trivial, since they coincide with prevailing expectation of experts. Such elections are identified by *Congressional Review.* Eliminating them from the score still results in 99% significance.

3.3 Presidential Elections

The prediction algorithm was developed by a retrospective analysis of the data on the past 31 elections, 1860 – 1980; that covers the period between victories of A. Lincoln and R. Reagan inclusively. The questionnaire is shown in Table 7. Victory for the challenger is predicted if the number of answers *no* is 6 or more (Lichtman, 1996, 2000).

Predicting of future elections. This algorithm (without any changes from year to year state) was applied by Prof. A. Lichtman in advance to the seven subsequent elections, 1984 – 2008 (Table 8). Predictions are shown in Table 9. All of them happened to be correct. In 2000 the decision of popular majority was reversed by the Electoral College; such reversals are not targeted by this algorithm (Lichtman, 1996, 2000).

3.4 Understanding Elections

Collective behavior. The finding that aggregate-level parameters can reliably anticipate the outcome of both presidential and senatorial elections points to an electoral behavior highly integrated not only for the nation as a whole but also within the diverse American states.

— A presidential election is determined by collective, integrated estimation of performance of incumbent administration during the previous four years.

— In case of senatorial elections the electorate has more diffused expectations of performance but puts more importance on political experience and status than in the case of presidential elections. Senate incumbents, unlike presidential ones, do not suffer from a bad economy or benefit from a good one. (This suggests that rather than punishing the party holding a Senate seat for hard times, the voters may instead regard the incumbent party as a safe port in a storm).

Similarity. For each election year in all states the outcomes of elections follow the same pattern that transcends the diversities of the situations in each of the individual elections.

The same pattern of the choice of the US President prevails since 1860, i.e. since election of A. Lincoln, despite all the overwhelming changes in the electorate, the economy, the social order and the technology of politics during these 130 years. (For example, the electorate of 1860 did not include some of the groups, which constitute 3/4 of present electorate, such as women, African Americans, or most of the citizens of the Latin American, South European, Eastern European, and Jewish descent (Lichtman, 2000).

An alternative (and more traditional) concept of American elections focuses on the division of voters into interest and attitudinal groups. By this concept the goal of the contestants is to attract maximum number of voting blocks with minimal antagonism from other blocks. Electoral choice depends strongly on the factors irrelevant to the essence of the electoral dilemma (e.g. on the campaign tactics). The drawbacks of this concept are discussed in (Lichtman, 2000; Keilis-Borok and Lichtman, 1993). In sum, the work on presidential and senatorial elections described above suggests the following new ways of understanding American politics and perhaps the politics of other societies as well.

Table 7. Questionnaire for Presidential elections (Lichtman, 1996, 2000)

KEY 1 (Party Mandate): After the midterm elections, the incumbent party holds more seats in the U.S. House of Representatives than it did after the previous midterm elections.

KEY 2 (Contest): There is no serious contest for the incumbent-party nomination.

KEY 3 (Incumbency): The incumbent-party candidate is the sitting president.

KEY 4 (Third party): There is no significant third-party or independent campaign.

KEY 5 (Short-term economy): The economy is not in recession during the election campaign.

KEY 6 (Long-term economy): Real per-capita economic growth during the term equals or exceeds mean growth during the previous two terms.

KEY 7 (Policy change): The incumbent administration effects major changes in national policy.

KEY 8 (Social unrest): There is no sustained social unrest during the term.

KEY 9 (Scandal): The incumbent administration is unattained by a major scandal.

KEY 10 (Foreign/military failure): The incumbent administration suffers no major failure in foreign or military affairs.

KEY 11 (Foreign/military success}: The incumbent administration achieves a major success in foreign or military affairs.

KEY 12 (Incumbent charisma): The incumbent-party candidate is charismatic or a national hero.

KEY 13 (Challenger charisma): The challenging-party candidate is not charismatic or a national hero.

0	1	2	3	4	5	6	7
			OK98				
			CO98				
			FL98				
			GA98				
			HA98	TN02			
			ID98	SC02			
			MA98	NC02			
			ND98	NE02			
			PN98	KY02			
			SD98	IA02			
			UT98	CO02			
			FL94	AL02			
			HA94	AK98			
			IN94	CA98			
			MT94	CT98			
			NB94	NE98			
			NJ94	OR98			
			TX94	SC98			
			WA94	VT98			
		AS98	WV94	WA98			
		KA98	WI94	CT94			
		LA98	AK90	MD94			
		MI98	IN90	NV94			
		NH98	KN90	WY94			
		MS94	ME90	CO90			
	AL98	NM94	MA90	HA90			
	AZ98	ND94	MT90	KY90			
	IO98	RI94	NB90	MI90			
	DL94	VT94	NC90	AZ86			
	MA94	AS90	TX90	CO86			
	NY94	IO90	WY90	ID86			
	AL90	MS90	AR86	LA86			
	DE90	NM90	CA86	NY86			
	IL90	OR90	IL86	OK86	WI98	MN94	
	LA90	RI90	IN86	WI86	CA94	MO94	
	OK90	SD90	IA86	NC86	ID90	VA94	
	SC90	VA90	NH86	WA86	PA86	NH90	
	TN90	WV90	OR86	MN90	IL98	IN98	
	HI86	AK86	VT86	OK94	ME94	OH98	
	OH86	CT86	TN94	PA94	AL86	MI94	
UT94	SC86	KS86	TX02	TN294	FL86	MD86	KY98
GA90	UT86	KY86	OK02	NC98	GA86	NV86	AZ94
NJ90	NH02	ND86	NJ02	NY98	MO86	SD86	OH94

| 0 | 1 | 2 | 3 | 4 | 5 | 6 | 7 |

OK98 – incumbent won, **KY98** – challenger won, errors are highlighted.

Figure 14 Made-in-advance predictions of the mid-term senatorial elections (1986-2002). Each election is represented by the two-letter state abbreviation with the election year shown by two last digits. Each column shows elections with certain number D of answers "*no*" to the questionnaire given in Table 6 (such answers are favourable to challenging party). Value of D, indicated at the top, is the Hamming distance from the kernel).

Table 8. Timing and source of predictions

Election	Date of prediction	Source
1984	APRIL 1982	"How to Bet in '84," *Washingtonian Magazine*, April 1982
1988	MAY 1988	"How to Bet in November," *Washingtonian Magazine*, May 1988
1992	SEPTEMBER 1992	"The Keys to the White House," *Montgomery Journal*, September 14, 1992
1996	OCTOBER 1996	"Who Will Be the Next President?" *Social Education*, October 1996
2000	NOVEMBER 1999	"The Keys to Election 2000" *Social Education*, November/December 1999.
2004	APRIL 2003	"The Keys to the White House," *Montgomery Gazette*, Apr. 25, 2003
2008	FEBRUARY 2006	"Forecast for 2008," *Foresight*, Feb. 2006,

An alternative (and more traditional) concept of American elections focuses on the division of voters into interest and attitudinal groups. By this concept the goal of the contestants is to attract maximum number of voting blocks with minimal antagonism from other blocks. Electoral choice depends strongly on the factors irrelevant to the essence of the electoral dilemma (e.g. on the campaign tactics). The drawbacks of this concept are discussed in (Lichtman, 2000; Keilis-Borok and Lichtman, 1993). In sum, the work on presidential and senatorial elections described above suggests the following new ways of understanding the American politics and perhaps the politics of other societies as well.

1. Fundamental shifts in the composition of the electorate, the technology of campaigning, the prevailing economic and social conditions, and the key issues of campaigns do not necessarily change the pragmatic basis on which voters choose their leaders.

2. It is the governing, not the campaigning, that counts in the outcomes of presidential elections.

3. Different factors may decide the outcome of executive as compared to legislative elections.

4. Conventional campaigning will not improve the prospects for candidates faced with an unfavorable combination of fundamental historical factors. Disadvantaged candidates have an incentive to adopt innovative campaigns that break the pattern of conventional politics.

5. All candidates would benefit from using campaigns to build a foundation for governing in the future.

Table 9. Prediction of US presidential elections (up to 2008)

Years when incumbent won are shown in italic; years when challenger won – in bold.

Number of keys in favor of a challenger party									
0	1	2	3	4	5	6	7	8	9
Predictions (published months in advance)									
				*2000**					
	1984	*1988*	*2004*	*1996*	**1992**			*2008*	
Retrospective analysis									
			1964					**1980**	
			1928					**1976**	
			1916					**1968**	
			1908					**1952**	
		1944	*1900*	*1972*				**1932**	
	1956	*1940*	*1872*	*1924*	*1948*	**1912**	**1884**	**1920**	**1960**
1904	*1936*	*1868*	*1864*	*1880*	**1888***	**1892**	**1860**	**1896**	**1876***

* years when popular vote was reversed by electoral vote.

CHAPTER 4

Prediction of Economic Recessions

The five recessions in the USA from 1961 to 1996 have been considered, and it has been found that each of them was preceded by a specific pattern of 6 economic indexes, which are defined at the lowest (binary) level of resolution (Keilis-Borok et al., 2000). This pattern was present during 6 to 14 month before each recession and at no other time, suggesting a hypothetical prediction algorithm. The algorithm is exceedingly robust: the retrospectively diagnosed alarms remain about the same after variation of its adjustable numerical parameters, and of other non-unique decisions involved in its determination. It has also been found that another robust pattern of the same six macroeconomic indicators appeared within 6 months before the ends of these recessions and at no other time during the recessions (Keilis-Borok et al., 2008). The last study is a natural continuation of the previous one, aimed at predicting the start of a recession. It is found by comparing these cases that precursory trends of financial indicators are opposite during the transition to a recession and the recovery from it. To the contrary, precursory trends of economic indicators happen to have the same direction (upward or downward), but are steeper during recovery.

4.1. Prediction Targets

US National Bureau of Economic Research (NBER) has identified seven recessions that occurred in the US from 1960 to 2002 (Table 10). The starting points of a recession and of the recovery from it follow the months marked by a peak and a trough of economic activity, respectively.

Table 10. American Economic Recessions, 1960-2002 (year:month)

#	Peaks	Troughs
1	1960:04	1961:02
2	1969:12	1970:11
3	1973:11	1975:03
4	1980:01	1980:07
5	1981:07	1982:11
6	1990:07	1991:03
7	2001:03	2001:11

A peak indicates the last month before a recession, and a trough — the last month of a recession.

Prediction targets considered are the first month after the peak and the first month after the trough ("the turns to the worst and to the best", respectively). The start of the first recession, in 1960, is not among the targets, since the data do not cover a sufficient period of time preceding the recession.

4.2. Data and Their Preliminary Analysis

The data used for prediction comprise the following six monthly leading economic indicators obtained from the CITIBASE data base (abbreviations are the same, as in Stock and Watson, 1993).

IP: Industrial Production, total: index of real (constant dollars, dimensionless) output in the entire economy. This represents mainly the manufacturing industry, because of the difficulties in measuring the quantity of the output in services (such as travel agents, banking, etc.).

INVMTQ: Total inventories in manufacturing and trade, in real dollars. Includes intermediate inventories (for example held by manufacturers, ready to be sent to retailers) and final goods inventories (goods on the shelves in stores).

LHELL: Index of "help wanted" advertising. This is put together by a private publishing company that measures the amount of job advertising (column-inches) in a number of major newspapers.

LUINC: Average weekly number of people claiming unemployment insurance.

FYGM3: Interest rate on 90 day U.S. treasury bills at an annual rate (in percent).

G10FF=FYGT10–FEDFUN: Difference between the annual interest rate on 10 year U.S. Treasury bonds, and federal fund annual interest rate.

These indicators were already known (Stock and Watson, 1989, 1993), as those that correlate with a recession's approach. The first four indicators concern the economy, while the two others concern the financial market.

Premonitory behavior of an indicator may be better captured by its linear trend.

Notations:

$f(m)$ is a time series of a monthly indicator, hypothetically containing a premonitory pattern; integer m is the time in months.

$W^f(l/q,p)$ – the local linear least-squares regression of a function $f(m)$ within the sliding time window (q, p):

$$W^f(l/q,p) = K^f(q,p)l + B^f(q,p), \quad q \le l \le p, \tag{1}$$

integers l, q, and p stand for time in months.

Two following transformations of indicators that reflect their trends are considered:

$$K^f(m/s) = K^f(m\text{-}s,m). \tag{2}$$

In notations of (1) this is the trend of $f(m)$ in the sliding time window s months long, $(m\text{-}s,\ m)$. It may be considered for prediction since its value is attributed to the *end* of window (m) and does not depend on the information about the future (after the month m) which would be anathema in prediction.

$$R^f(m/q) = f(m) - W^f(m/q,m\text{-}1). \tag{3}$$

This is deviation of an index from its long-term extrapolation (since long time window $(q, m - 1)$ is considered in estimating W^l).

Discretization. Functions used in a prediction algorithm are considered on the lowest – binary – level of resolution, as 0 or 1, distinguishing only the values above ("large") and below ("small") a threshold T. Then the months (objects of recognition) are described by binary vectors of the same length. Discretization ensures robustness of analysis. The threshold is defined as a Q – level percentile, e.g. by the condition that the function F exceeds $T^F(Q)$ during $Q\%$ of the months considered.

4.3. Prediction of a Recession Start

Time periods. The time period from 1960 to 1996 was considered by Keilis-Borok et al. (2000) when the prediction algorithm was designed. The time of each recession and 5 months after it is eliminated from consideration, since the behavior of the indexes may have some special features during a recession and its aftermath; the interval of 5 months is chosen, because subsequent recessions are officially considered as different ones only if they are separated by at least six months. At this stage, the period before the first recession in Table 10 is also excluded (since it starts too close to the beginning of the data set), as well as the period after the sixth one. Thus, the following 5 periods are used for the search of premonitory phenomena:

W_1: 1961:08 - 1969:12 (101 months);
W_2: 1971:05 - 1973:11 (31 months);
W_3: 1975:09 - 1980:01 (53 months);
W_4: 1981:01 - 1981:07 (7 months);
W_5: 1983:05 - 1990:07 (87 months).

The combination of these periods $W = W_1 \cup W_2 \cup W_3 \cup W_4 \cup W_5$ contains 279 months. These months between recessions are considered when the discretization thresholds $T^F(Q)$ are determined.

Indicators describing the months are transformed into functions given in Table 11. By means of discretization with the values of Q given in Table 10 the functional values are reduced to "small" or "large".

The goal of the algorithm is to predict the beginning of a recession, defined, as the first month after the "peak" (table 10). Prediction algorithm, considered here, has a limited accuracy; it may indicate not a single month, but a time interval several months long, which includes the first month of a recession. Continues intervals of this kind are called "alarms". A recession is regarded as being "predicted", if it begins within an alarm and as a "failure to predict" in the opposite case. A "false alarm" is an alarm during which no recession has started.

Table 11. Functions used for the indicators and their typical values in pre-recession periods

Indicator	Function used for the indicator	Q, %	Typical value in pre-recession situation
IP	Deviation from the long-term trend $R^{IP}(m/q)$* (3)	75	Small
INVMTQ	Deviation from the long-term trend $R^{INVMTQ}(m/q)$* (3)	25	Small
LHELL	Short trend $K^{LHELL}(m/5)$ (2)	66.7	Small
LUINC	Short trend $K^{LUINC}(m/10)$ (2)	16.7	Large
FYGM3	Deviation from the long-term trend $R^{FYGM3}(m/q)$* (3)	25	Large
G10FF	Indicator itself	90	Small

*) q in the determination of $R^f(m/q)$ is the first month after the end (the through) of the previous recession

The whole set of months under consideration W is divided into two parts: the set W_D that includes three months before each recession, and the set $W_N = W \setminus W_D$. The value of a functional is typical for the set W_D (pre-recession situation) if a percent of months from W_D with the functional having this value is greater than this percent of months from W_N. The typical values of the functions in pre-recession situation are given in the last column of

Table 11. After this analysis all months are described by binary vectors with 6 components corresponding to the functionals listed in Table 11. Each component has been defined in such a way, that the value "1" would become typical in a pre-recession situation (more frequent when a recession is approaching). Accordingly, the description of pre-recession situations would be close to the vector $(1,1,1,1,1,1)$, which is called the kernel. Let $\Delta(m)$ be the number of zeros in a description of the month m - its Hamming's distance from the kernel. The approach of a recession may be recognised by the small values of $\Delta(m)$. A priori this is not clear, in spite of the way the zeros are defined; for example, premonitory changes of the functions may appear not simultaneously.

Prediction algorithm. The prediction algorithm is formulated as follows (Keilis-Borok et al., 2000): *an alarm is declared for three months after each month m with $\Delta(m) \leq 2$ (regardless of whether or not this month belongs to an alarm which has already been declared).*

Alarms and recessions are juxtaposed in Fig. 15. One can see that five recessions occurring between 1962 and 2000 (#2 – 6 in Table 10) are predicted by an alarm. Duration of each alarm is between 6 and 14 months. Total duration of all alarms is 43 months, or 15.4 % of the time interval (set W) considered. There are no false alarms. Months between two last recessions in Fig. 15 were not included in the set W and were not considered in the analysis described above. The last recession in Fig. 15 (No. 7 in Table 10) is preceded by an alarm determined by the algorithm.

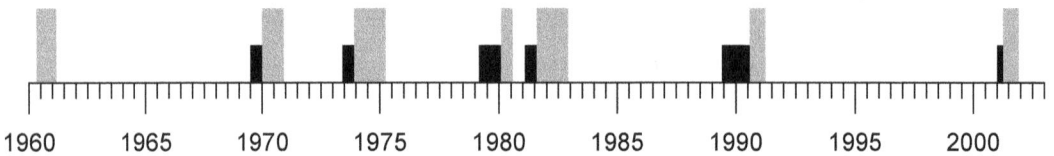

Figure 15. Alarms (black bars) and recessions (grey bars).

4.4. Prediction of Recovery from a Recession

The study on predictions of recovery from a recession (Keilis-Borok et al., 2008) is a natural continuation of the study aimed at predicting the start of a recession.

Prediction targets are defined above (see Sect. 4.1).

The data are the same six indicators that indicate the approach of a recession (see Sect. 4.2); they are analysed only within the recession period.

The time period from 1960:01 to 2002:04 is considered, when seven recessions did occur (Table 10). At the time of the study by Keilis-Borok et al. (2008) the trough for the last recession had not been yet established by the NBER and the data concerning only the first 6 recessions were used. Thus the following 6 periods are considered (for each recession a month before it (the peak) and its last month (the trough) are included).

W_1: 1960:04 - 1961:02 (11 months);
W_2: 1969:12 - 1970:11 (12 months);
W_3: 1973:11 - 1975:03 (17 months);
W_4: 1980:01 - 1980:07 (7 months);
W_5: 1981:07 - 1982:11 (17 months);
W_6: 1990:07 - 1991:03 (9 months)

The total duration of these periods W = $W_1 \cup W_2 \cup W_3 \cup W_4 \cup W_5 \cup W_6$ is 73 months. The last recession is considered separately, in Section 4.6.

As in Keilis-Borok et al. (2000), we use for prediction the values of the indicator **G1OFF**; the trends of the indicators **LHELL** and **LUINC**; and deviations from the long-term trend of the indicators **IP**, **INVMTQ**, and **FYGM3** (Table 12).

Analyzing the histograms of the six functionals considered during the recession periods Keilis-Borok et al. (2008) have found that large values of **G1OFF** and $K^{\text{LUINC}}(m/10)$ and small values of other functions occur more frequently as the termination of a recession approaches. To check whether this impression is correct, a robust quantitative definition of these changes is given by means of the discretization procedure described in Section 4.2. Analyzing behavior of each function, the values of Q indicated in Table 12 and the relevant values of $T^F(Q)$ have been determined.

The last column of Table 12 presents "precursory potential" for each function, i.e., the difference $P_e - P_b$, where P_e is percent of months with "precursory" value of a function in a set consisting of the last four months of each recession; P_b is percent of such months in a whole set W with the last six months of each recession eliminated. Looking at these data one can conclude that some indicators (e.g., $K^{LUINC}(m/10)$) may be used individually for determination of the precursory pattern. However the existing experience shows that using several indicators makes the result more reliable.

Table 12. Functions used for the indicators and their typical values before recovery from a recession

Indicator	Function used for the indicator	Q, %	Typical value before recovery from a recession	"Precursory potential"**, %
IP	Deviation from the long-term trend $R^{IP}(m/q)$* (3)	75	Small	67
INVMT Q	Deviation from the long-term trend $R^{INVMTQ}(m/q)$* (3)	50	Small	20
LHELL	Short trend $K^{LHELL}(m/5)$ (2)	75	Small	45
LUINC	Short trend $K^{LUINC}(m/10)$ (2)	50	Large	70
FYGM3	Deviation from the long-term trend $R^{FYGM3}(m/q)$* (3)	50	Small	20
G10FF	Indicator itself	33.3	Large	67

*) q in the determination of $R^f(m/q)$ is the first month after the end (the through) of the previous recession. An exception is the first recession for which q is the first month of the time period considered, 1960:01 (since the previous recession ends before this period).

**) Difference $P_e - P_b$ where P_e is percent of months with "precursory" value of a function in a set consisting of the last four months of each recession; P_b is percent of such months in a whole set W with the last six months of each recession eliminated.

After discretization of the functions, the situation during the recession is robustly described in relation to its proximity to a recession's end. This description is reduced to a monthly time series of a binary vector with 6 components. For convenience, the same code, "1", is given to the "precursory" trend of each function, regardless of whether it is above or below the threshold. Accordingly, the code "1" is given to the large values of the functions **G10FF** and $K^{\text{LUINC}}(m/10)$, and to the small values of the four other functions. Other values receive code "0".

The precursory pattern is looked for that indicates that a recession will end within a few months after this pattern appears. Continuous intervals of this kind are called "alarms." Possible outcomes of prediction based on this pattern are illustrated in Fig. 16. The probabilistic nature of the prediction is reflected in probabilities of the errors of each kind and in the percent of the time occupied by the alarms. This formulation of the prediction problem is discussed by Molchan (1997); it was applied by Keilis-Borok et al. (2000) for predicting of the start of a recession.

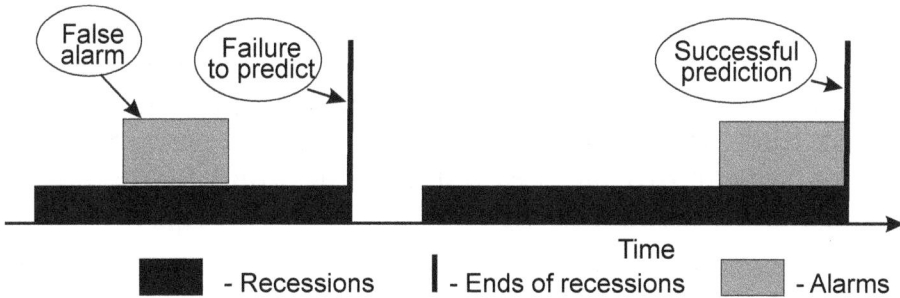

Figure 16. Outcomes of prediction of recovery from a recession.

After discretization, the monthly description of a situation during a recession is reduced to a binary vector with 6 components. Each component has been defined in such a way, that the values "1" would became more frequent when the end of a recession is approaching. Accordingly, the "ideal" situation prior to the end of a recession, when all indicators are precursory, is the vector (1,1,1,1,1,1), which is called the "kernel." Let $\Delta(m)$ be the number of zeros in a code of a month m – its Hamming distance from the kernel. A question is, whether the approach of the end of a recession may be recognized by the small values of $\Delta(m)$. A priori this is not clear, despite the way the kernel is defined; for example, precursory changes of the functions may not appear simultaneously.

Prediction algorithm. Keilis-Borok et al. (2008) have found that the end of each recession is preceded by a continuous group of months with Δ(m) ≤ 3. This suggests the following algorithm: *the precursory pattern appears if for three consecutive months Δ(m) ≤ 3 and the recession end is expected during an interval of three months long after displaying this pattern.*

The alarms (continuous intervals of this kind) obtained by this algorithm for the recessions considered are shown in Fig. 17. The end of each of the 6 recessions is preceded by an alarm and there are no false alarms. The duration of an alarm inside each recession is given in Table 13. The total duration of alarms in all 6 recessions is 16 months, which is 22% of total duration of the set *W* (73 months).

The duration of an alarm does not exceed 5 months; this means that the precursory pattern appears during the last 6 months of each recession. Let us estimate the probability *P* to obtain such result by chance, when the pattern appears independently on when recession ends, at a random moment uniformly distributed over the recession period. Taking into account that, according to its definition, the precursory pattern may not appear during the first two months of a recession, the value of *P* is calculated as follows

$$P = \frac{6}{9} \times \frac{6}{10} \times \frac{6}{15} \times 1 \times \frac{6}{15} \times \frac{6}{7} = 0.055 .$$

Here the denominators are the durations of the recessions (Table 10) reduced by 2. The factor corresponding to the fourth recession is 1 because it is so short that the alarm for its end may not be longer than 5 moths. The probability is encouragingly low, although additional out-of-sample tests are necessary to establish statistical significance (Molchan et al., 1990).

Table 13. Alarms before the ends of the recessions

Recession	Trough	The first month of an alarm and its duration (in months) before the recession end
1	1961:02	1960:12, 3
2	1970:11	1970:08, 4
3	1975:03	1975:02, 2

4	1980:07	1980:07,	1
5	1982:11	1982:07,	5
6	1991:03	1991:03,	1
7	2001:11	2001:07,	5

Recessions are numbered as in Table 10.

The approach of recession is captured (Section 4.3) by the same indicators that are used here to capture recovery from it. Behavior of the indicators during these opposite processes is compared in Table 14 and Fig. 18. One can see reverse trends of "financial" indicators; and, contrary to what might be expected, the trends of "economic" indicators have the same direction, but are steeper during recovery.

4.5 Sensitivity Analysis

The analysis given in the two previous sections, inevitably for such a heuristic one, does involve considerable data fitting, e.g. the choice of the functions, numerical parameters, etc. (Dr. W. Press, discussing an early example of such an approach, reminded the words of E. Fermi: "With four exponentials I can fit an elephant"). Therefore it is necessary to explore how stable are the results to variation of these choices. If they are sufficiently stable, one may hypothesize that the same patterns will precede the starts of the subsequent recessions and the recovery from them. The only decisive test of this hypothesis could be provided by advance predictions.

The following tests were made by Keilis-Borok at al. (2000) to check the stability of the algorithm that predicts **a recession start**. Its application is considered below as "main variant".

"*Prediction history*". The discretization thresholds $T^F(Q)$ used by the algorithm are determined considering the whole set W (Section 4.3). In the test the thresholds $T^F(Q)$ are determined for consecutively expanding time periods ("T-sets"): W_1, $W_1 \cup W_2$, $W_1 \cup W_2 \cup W_3$, and $W_1 \cup W_2 \cup W_3 \cup W_4$. The values of $T^F(Q)$ for each period are calculated for the same values of Q that are given in Table 12. The prediction algorithm with each set of thresholds was applied to the subsequent time period up to a next recession, that is, to the sets W_2, W_3, W_4, and W_5 respectively. The alarms obtained in the experiment are shown in Fig. 19b. The alarms obtained in the main variant are shown for comparison in Fig. 19a. The agreement with the main variant is practically complete. The only difference is that the last alarm became a month longer.

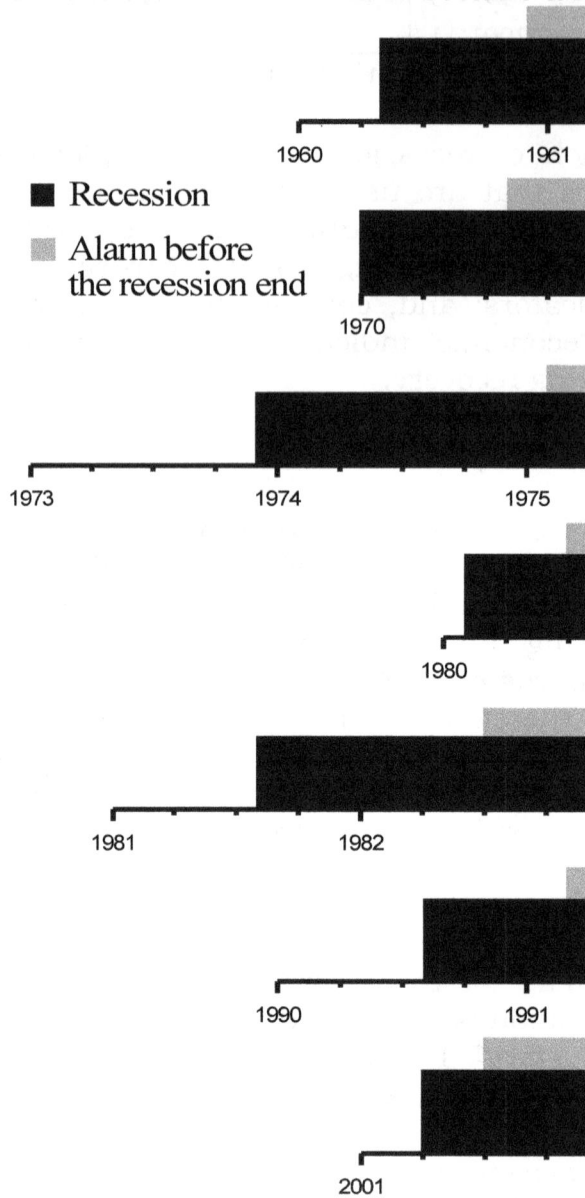

Figure 17. Results of application of the algorithm predicting recovery from a recession.

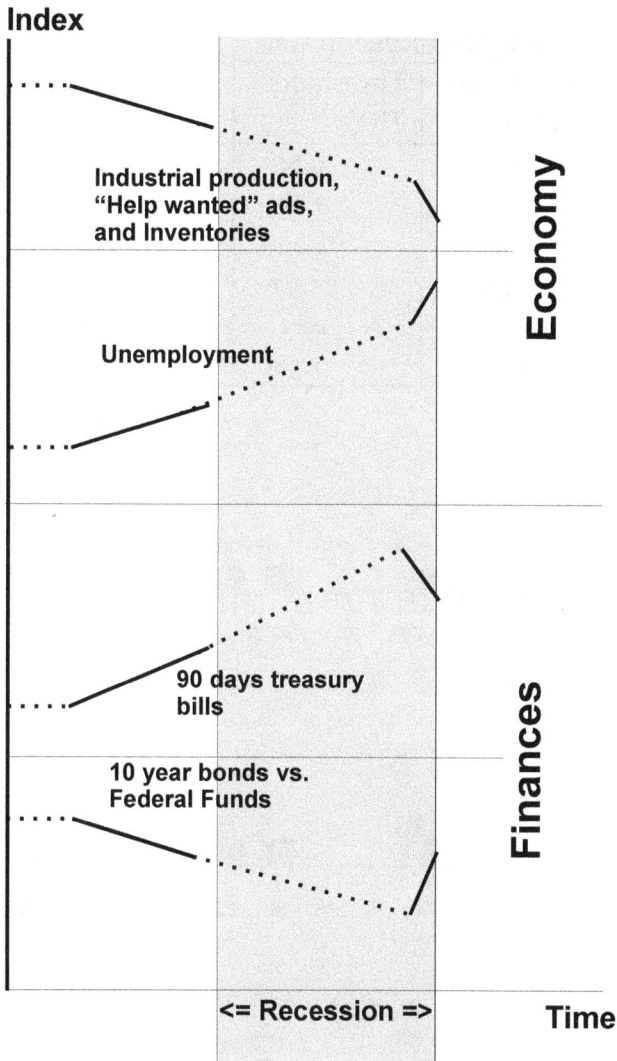

Figure 18. Premonitory changes of indicators before the start of a recession and before its end.

Table 14. Precursory behavior of indicators

Function	Before the recession start		Before the recession end	
	Precursory values	Threshold $T^F(Q)$	Precursory values	Threshold $T^F(Q)$
$R^{IP}(m/q)$	low	-0.86	low	-3.71
$R^{INVMTQ}(m/q)$	low	8.85	low	6.50
$K^{LHELL}(m/5)$	low	2.00	low	-29.71
$K^{LUINC}(m/10)$	high	25.07	high	112.15
$R^{FYGM3}(m/q)$	high	1.01	low	-0.64
G10FF	low	-1.38	high	0.82

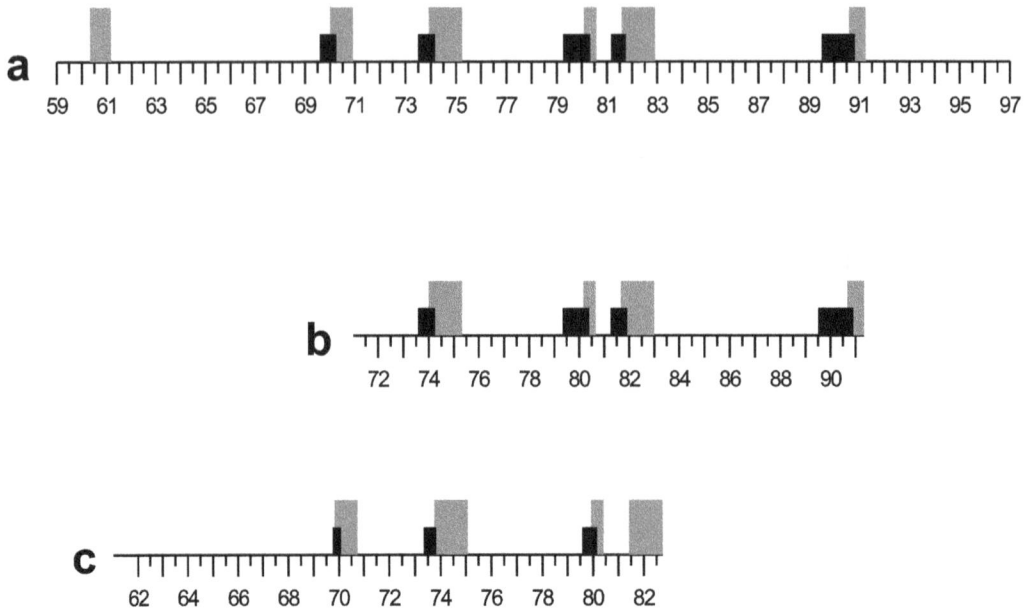

Figure 19. Alarms (shown by black bars): a - main version; b - prediction history; c - reverse prediction history.

"Reverse prediction history". In this test, the data set is expanded in the reverse order: the thresholds $T^F(Q)$ are determined consecutively for the T-sets W_5, $W_5 \cup W_4$, $W_5 \cup W_4 \cup W_3$,

and $W_5 \cup W_4 \cup W_3 \cup W_2$ and the prediction algorithm is applied to the sets W_4, W_3, W_2, and W_1 respectively. The alarms obtained in this test are shown in Fig. 19c. One can see that the recession in 1981 - 1982 is missed by the prediction; the alarms before the recessions in 1969 - 1970 and in 1980 are shorter than in the main variant.

"Change of discretization". In this test the values of Q listed in Table 11 are increased and decreased as shown in Table 15. Again, the results are similar to those in the main variant, though the changes are not insignificant: one false alarm appeared; three alarms of the main version became separated into two parts, so that in statistical tests the first parts should be counted as false alarms.

Table 15. Values of Q that were used when discretization was changed in the algorithm of recession start prediction (Keilis-Borok et al., 2000)

Function	Q in the main variant, %	Q in the test "Change of discretization", %	
$R^{IP}(m/q)$	75	66.7	80
$R^{INVMTQ}(m/q)$	25	20	33.3
$K^{LHELL}(m/5)$	66.7	50	75
$K^{LUINC}(m/10)$	16.7	14.3	20
$R^{FYGM3}(m/q)$	25	20	33.3
G10FF	90	87.5	92

"Elimination of the functions". In this test, each of the six functions used in the algorithm is eliminated from consideration, one at a time. The alarms are declared when $\Delta(m) \leq 1$ instead of 2 in the main variant. The major change with respect to the main variant is failure to predict the recession in 1981 - 1982 after either of the functions $R^{IP}(m/q)$, $R^{INVMTQ}(m/q)$, $K^{LHELL}(m/5)$, or **G10FF** is eliminated. The smallest changes are caused by elimination of $K^{LUINC}(m/10)$. When either of the functions $R^{INVMTQ}(m/q)$, $K^{LHELL}(m/5)$, $R^{FYGM3}(m/q)$, or **G10FF** is eliminated, the alarm before the recession in 1973 - 1975 starts in the same month as the recession.

"Prediction by a single function". In this test, each of the functions is used separately of others. Obviously, the alarms were declared if the value of the relevant binary component is 1. When the function $R^{IP}(m/q)$, $K^{LUINC}(m/10)$, $R^{FYGM3}(m/q)$, or **G10FF** is used, there is one failure to predict; total duration of alarms is the smallest with **G10FF**, the largest with $R^{IP}(m/q)$, and about the same with $K^{LUINC}(m/10)$ and $R^{FYGM3}(m/q)$. When $R^{IP}(m/q)$ or $K^{LHELL}(m/5)$ is used, total duration of alarms is much larger, but there are no false alarms.

Performance of different of predictions obtained in the tests is juxtaposed in the error diagram shown in Fig. 20.

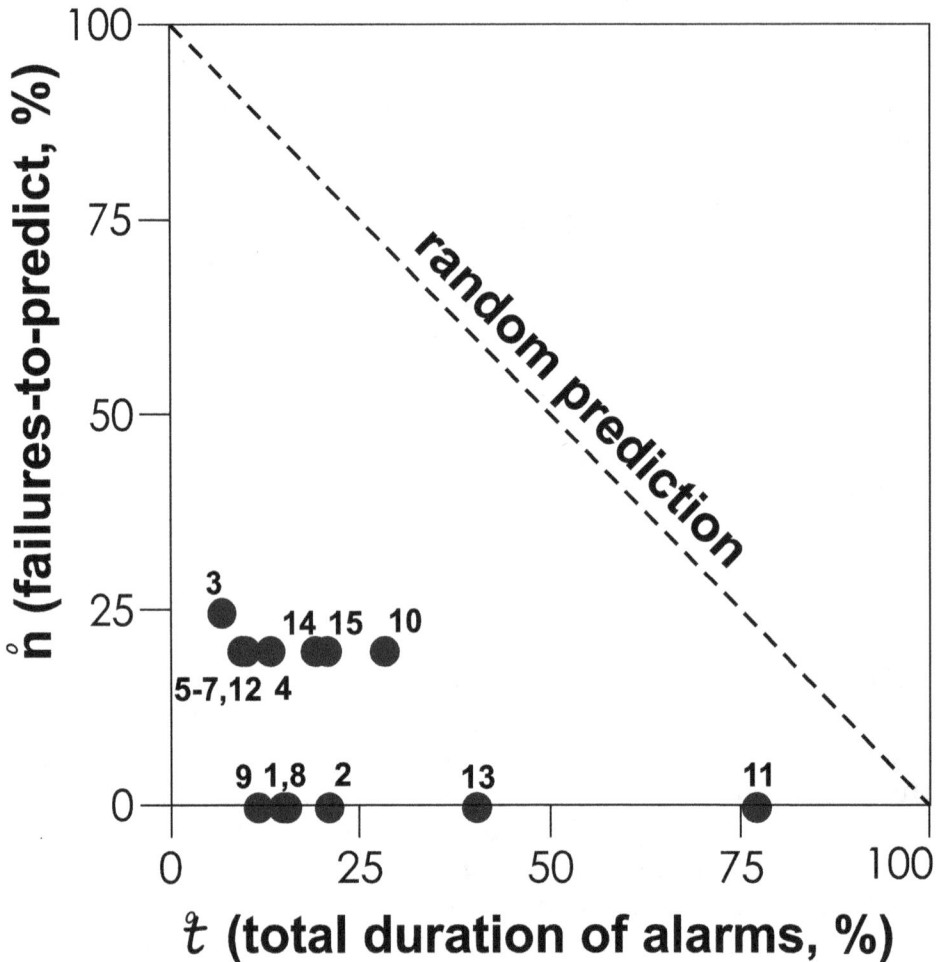

Figure 20. Error diagram: 1 - main version; 2 - prediction history; 3 - reverse prediction history; 4-9 exclusion of functions: $R^{IP}(m/q)$ (4), $R^{INVMTQ}(m/q)$ (5),

G10FF (6), $K^{LHELL}(m/5)$ (7), $K^{LUINC}(m/10)$ (8), $R^{FYGM3}(m/q)$ (9); 10-15 - prediction by a single function: $R^{IP}(m/q)$ (10), $R^{INVMTQ}(m/q)$ (11), G10FF (12), $K^{LHELL}(m/5)$ (13), $K^{LUINC}(m/10)$ (14), $R^{FYGM3}(m/q)$ (15).

The following tests were made by Keilis-Borok at al. (2008) to check the stability of the algorithm that predicts **recovery from a recession**. Its application is considered below as "main variant".

Table 16. Changing of the alarms given in Table 13 when the functions are removed

Recession	Functions removed and the first month of alarms obtained and its duration (in months) before the recession end ("–" indicates cases where there is no difference with the result given in Table 13)					
	R^{IP} (m/q)	R^{INVMTQ} (m/q)	K^{LHELL} $(m/5)$	K^{LUINC} $(m/10)$	$R^{FYGM3}(m/q)$	G10FF
1	–	No alarm		No alarm	1961: 01, 2	No alarm
2	1970: 10, 2	1970: 10, 2		1970: 10, 2	–	–
3	–	–	–	–	–	–
4	1980: 08, 0	–	1970: 09, 3	1980: 08, 0	–	–
5	–	1982: 10, 2	–	1982: 10, 2	1982: 10, 2	–
6	–	–	1980: 08, 0	–	–	–

Recessions are numbered as in Table 10.

Elimination of the functions. As above, it is studied in this test how the prediction changes if one of the functions is removed from consideration. For that purpose, each of the 6 functions used in the prediction algorithm (Section 4.4) is removed in turn. Taking into account that in this case the length of the binary codes of months reduces to 5 the condition $\Delta(m) \leq 3$ in the algorithm formulated in Section 4.4 is changed by the condition $\Delta(m) \leq 2$. The results of this experiment are given in Table 16. The test shows stability of the prediction results to excluding of functions. Only for the first recession (1960-1961) its end is failure-to-predict when the function $R^{INVMTQ}(m/q)$, or $K^{LUINC}(m/10)$ or **G10FF** is removed. In other cases there are only some variations in durations of alarms. The greatest number of differences is in the case where the function $K^{LUINC}(m/10)$ is removed that is in accordance with the largest "precursory potential" of this function (see Table 12).

Table 17. Changing of the alarms given in Table 13 when the indicators are used individually

Recession	Functions used individually and the first month of alarms obtained and its duration (in months) before the recession end ("–" indicates cases where there is no difference with the result given in Table 13)					
	R^{IP} (m/q)	R^{INVMTQ} (m/q)	K^{LHELL} (m/5)	K^{LUINC} (m/10)	R^{FYGM3} (m/q)	G10FF
1	No alarm	1960:07, 8	No alarm	1961:01, 2	1960:07, 7 (false alarm)	–
2	1970:09, 3	–	No alarm	1970:07, 5	No alarm	1970:11, 1
3	–	No alarm	1975:01, 3	–	1975:01, 3	No alarm
4	–	No alarm	1980:06, 2	–	1980:08, 0	No alarm
5	No alarm	1981:10, 6 (false alarm) 1982:07, 5	No alarm	1982:03, 9	1981:10, 14	1982:10, 2
6	–	No alarm	1991:02, 2	–	No alarm	1991:04, 0

Recessions are numbered as in Table 10.

Prediction by a single function. The purpose of this test is to know what results would be obtained if the functions are used individually in the prediction algorithm. In this case the length of binary codes of months is 1 and the condition $\Delta(m) \leq 3$ in the algorithm formulated in Section 4.4 is changed by the condition $\Delta(m) = 0$. The results of the individual use of functions are given in Table 17. One can see that the functions used individually do not ensure good quality of prediction. Only using of $K^{LUINC}(m/10)$ does not result in false alarms or failures-to-predict, but the alarm before the end of the fifth recession (1981-1982) is too long.

Variation of the set of months used in determination of thresholds $T^F(Q)$. The thresholds of discretization $T^F(Q)$ have been determined by using a set $W = W_1 \cup W_2 \cup W_3 \cup W_4 \cup W_5 \cup W_6$ (Section

4.4) containing the months belonging to the 6 recessions under consideration. To check the sensitivity of the result to the composition of a training set the thresholds $T^F(Q)$ are computed excluding in turn the subsets W_i (i = 1, 2, ..., 6) from the training set W. The results of predictions with different values of $T^F(Q)$ are given in Table 18. These results show that the prediction is stable with respect to varying the set of months used in determination of the discretization thresholds. There is only one failure-to-predict case (the end of the second recession when the months of the third recession are excluded from the training set) and there are no false alarms.

Table 18. Changing of the alarms given in Table 13 when different training sets are used in determination of the thresholds $T^F(Q)$

Recession	Training sets and the first month of alarms obtained and its duration (in months) before the recession end ("–" indicates cases where there is no difference with the result given in Table 13)					
	$W\backslash W_1$	$W\backslash W_2$	$W\backslash W_3$	$W\backslash W_4$	$W\backslash W_5$	$W\backslash W_6$
1	1960:09, 6	1961:01, 2	1961: 01, 2	–	1960:11, 2	1961:01, 2
2	1970:11, 1	1970:07, 5	No alarm	–	1970:10, 2	–
3	–	–	–	–	–	–
4	–	–	–	–	–	–
5	1982:06, 6	1982:06, 6	–	–	1982:06, 6	–
6	–	–	–	–	–	–

Recessions are numbered as in Table 10.

Variation of parameter s in functions $K^{\text{LHELL}}(m/s)$ *and* $K^{\text{LUINC}}(m/s)$. The values of parameter s in functions $K^{\text{LHELL}}(m/s)$ and $K^{\text{LUINC}}(m/s)$ have been selected to be the same as those used in the algorithm for prediction of the start of a recession (Section 4.3): s = 5 and s = 10 respectively. These values are changed to see how the prediction results depend on them. When the function $K^{\text{LHELL}}(m/s)$ is calculated with s = 4 and s = 6, the alarms obtained by the algorithm are exactly the same as those given in Table 13. When the function $K^{\text{LUINC}}(m/s)$ is calculated with s = 8, there is only one difference in respect to Table 13: the first month of the alarm for the end of the first recession is 1960:11. If s = 8 for this function, then there are two differences: the first month of the alarm for the end of the first recession is 1961:03 and the first month of the alarm for the end of the third recession is 1975:01. Therefore the prediction is stable with respect to varying the value of *s*.

4.6. Experience in Advance Forecasting

The prediction algorithms described above were applied for advance forecasting after a recession occurred in 2001 (line 7 in Table 10). One recession had occurred during this period. Its first (January 2008) and last (June 2009) months were announced by NBER on 2008/12/01 and 2010/09/20, respectively. The application of the algorithm is shown in Fig. 21.

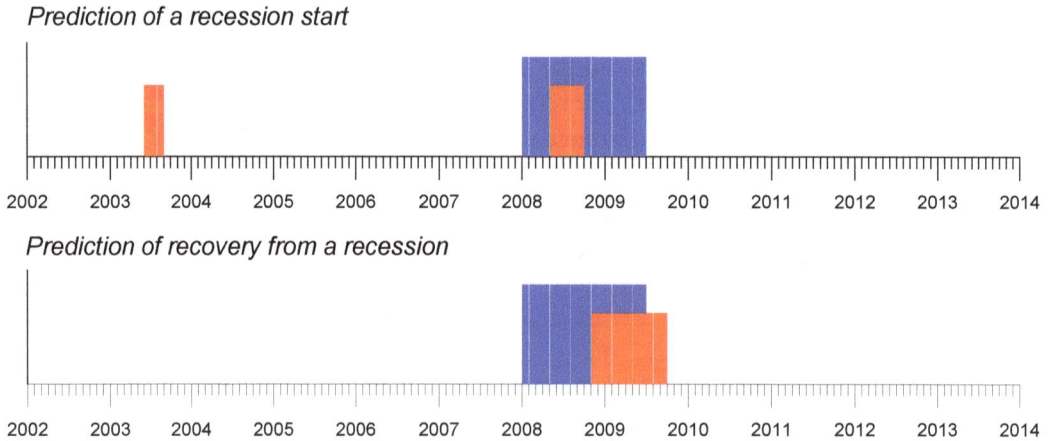

Figure 21. Advance forecasting. Blue bars show the period of the recessions, red bars – the periods of the relevant alarms.

Prediction of a recession start produced one false alarm in 2003 and the alarm on May 2008, four months later than the recession started, but six months before the announcement of its start by NBER.

An alarm for recovery from the recession started in November 2008, seven months before its last month. According to NBER the last recession ended in June 2009; this was established more than three months later. This illustrates an interesting aspect of the advance prediction. If the conclusions given above are correct, the algorithm indicates a time interval when a recession will end. However, testing the algorithm, one might not know when a recession actually ended, as that is decided later by the NBER after the comprehensive analysis. Not knowing whether the recession is still continuing, one would keep using the score Δ (applicable only within recessions) and thus might extend the alarm far into the inter-recession period. A heuristic way to overcome this difficulty could be suggested: since previous alarms lasted not more than 8 months (3.6 months on average), one can tentatively assume that an alarm will not last for more than, say, 9 months and put a 9 months limit for the duration of an alarm. An interesting question is whether the findings of this study (e.g. shown in Fig. 18) might help to recognize that a recession has ended.

CHAPTER 5

Prediction of Unemployment Surges

The study by Keilis-Borok et al. (2005) integrates the expertise in the fields of economics, labor relations, and critical phenomena in complex systems. A specific phenomenon in unemployment dynamics has been considered: episodes of a sharp increase in the unemployment rate called **"Fast Acceleration of Unemployment" or "FAU."** The study is a "technical" analysis that is a heuristic search of phenomena preceding *FAUs*. The methodology of *pattern recognition of infrequent events* has been used. It was developed by the artificial intelligence school of I.M. Gelfand for a study of rare phenomena of a highly complex origin, that, by their nature, limit the possibilities of using classical statistical or econometric methods. The goal of Keilis-Borok et al. (2005) was to identify by an analysis of macroeconomic indicators a robust and rigidly defined prediction algorithm of the *"yes or no"* variety indicating at any time moment, whether *FAU* should be expected or not within the subsequent *T* months. Considering unemployment in France between 1962 and 1997, a specific "premonitory" pattern of three macroeconomic indicators has been found that may be used for algorithmic prediction of *FAUs*: among seven *FAUs* identified within these years, six are preceded within 12 months by this pattern that appears at no other time. Application of that algorithm to Germany, Italy and the United States yields similar results. Such predictability reflects the fact that the economy, like many other complex systems, exhibits regular collective behavior patterns. The final test in any prediction research is advance prediction. Such predictions, for USA, have been successful.

5.1 Analysis of Data. Definition of Prediction Target

The problem: prediction of _FAUs_. Keilis-Borok et al. (2005) have considered a specific phenomenon in the dynamics of unemployment: a sharp increase in the rate of unemployment growth. Qualitatively this phenomenon is illustrated in Fig. 22. The thin line is the monthly number of unemployed $u(t)$, including seasonal variations. After smoothing $u(t)$ to eliminate these variations the function $U(t)$ is obtained. The target of the prediction is the starting time of a strong and lasting increase in the unemployment rate $U(t)$. An example is the turning point indicated by the arrow in Fig. 22. Keilis-Borok et al. (2005) termed this target _FAU_, for "Fast Acceleration of Unemployment." The goal is to design an algorithm for predicting the _FAU_s by an analysis of macroeconomic indicators.

Such an algorithm, if found and validated, may be useful in two ways: (i) As a quantitative and reproducible description of phenomena that are premonitory to _FAU_s; this would provide empirical constraints for the theoretical modeling of unemployment. (ii) As a practical tool, complementing existing methods of predicting unemployment. These are the usual twofold goals in any prediction research. Keilis-Borok et al. (2005) have developed such an algorithm for France (Section 5.2) and tested it on independent data for Germany, Italy, and USA (Section 5.3).

Methodology. This study belongs to the so-called "technical" analysis, consisting of a heuristic search for phenomena preceding _FAU_s. The alternative would be a "fundamental" analysis, focusing on "cause-and-effect" mechanisms leading to a _FAU_. The methodology that is used for technical analysis is the _pattern recognition of infrequent events_. It was developed by the artificial intelligence school of I.M. Gelfand (Gelfand et al., 1976) for the study of rare phenomena of highly complex origin. This approach differs from, but complements, classical statistical and econometric methods, such as regression analysis and ARIMA (Engle and McFadden, eds, 1994; see also Stock and Watson 1989; Klein and Niemira 1994; Mostaghimi and Rezayat 1996). For comparison of the pattern recognition with the multiple regression, see the paper by Keilis-Borok et al. (2000), on prediction of economic recessions. This methodology has been successfully applied also to predictions of the outcome of

American elections (Lichtman and Keilis-Borok, 1989; Keilis-Borok and Lichtman, 1993), as well as in seismology (Gelfand et al., 1976; Press and Briggs, 1975; Keilis-Borok and Press, 1980; Press and Allen, 1995; Keilis-Borok and Soloviev, 2003), geological prospecting (Press and Briggs, 1977), and many other fields, as given in the references in these papers. Here, the simplest version of this methodology is used, that is called the "Hamming distance" (Gvishiani and Kossobokov, 1981; Lichtman and Keilis-Borok, 1989; Keilis-Borok and Soloviev, 2003; and references therein). Its essence will be clear from the way in which the data are analyzed here.

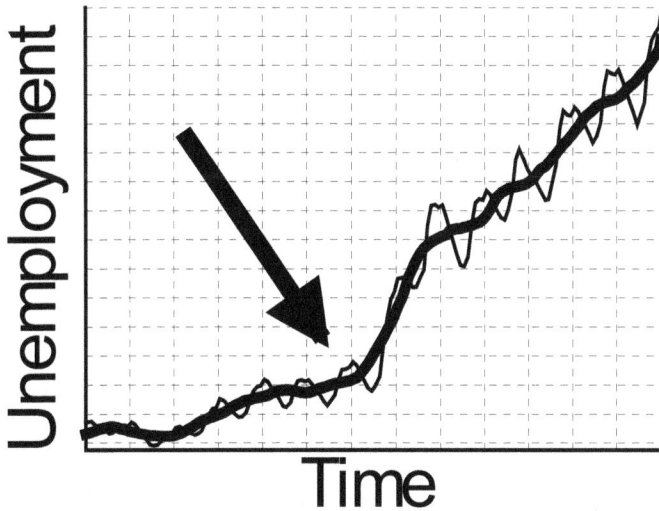

Figure 22. Fast acceleration of unemployment (*FAU*): schematic definition. Thin line – monthly unemployment; with seasonal variations. Thick line – monthly unemployment, with seasonal variations smoothed away. The arrow indicates a *FAU* – the sharp bend of the smoothed curve. The moment of a *FAU* is the target of prediction.

The specific features of this approach might be summarized as follows:

(i) It is applied to predicting not the whole dynamics of unemployment but only the relatively rare extraordinary phenomena - the *FAU*s.

(ii) A quantitative and precisely defined prediction algorithm of the *"yes or no"* variety is looked for, where, at any moment of time, the algorithm would indicate whether or not a *FAU* should be expected within the subsequent r months. The possible outcomes of such a prediction are illustrated in Fig. 23. The probabilistic nature of such prediction is reflected in the estimation of the probabilities of false alarms, failures to predict, and of the relative time occupied by alarms, as discussed below.

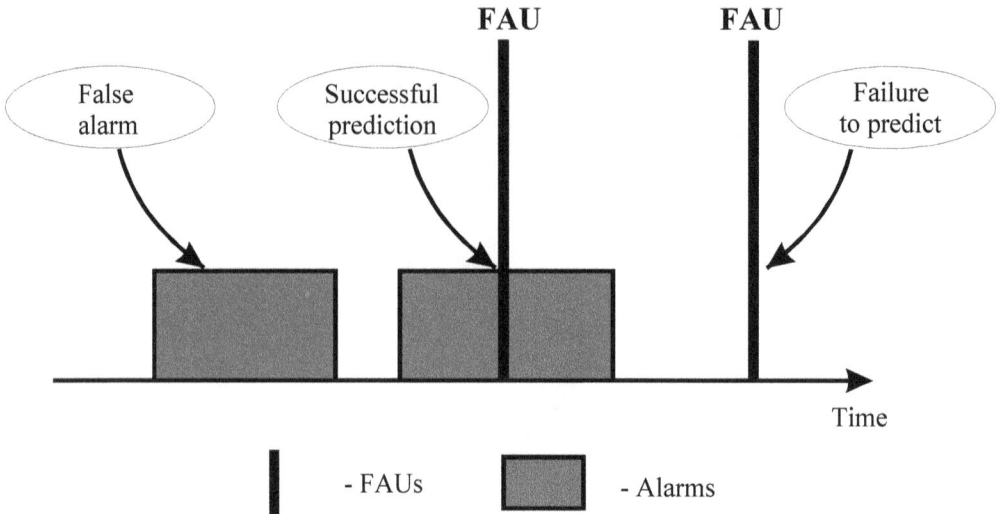

Figure 23. Possible outcomes of prediction.

(iii) The analysis is intentionally robust, which makes the prediction algorithm more reliable and applicable to different *FAUs*, despite the diversity of their specific causes. This is achieved at a price, however, in that the exactness of prediction is limited.

(iv) The analysis does not compete with, but rather complements the short-term predictions of the *"cause and effect"* kind. The cause that triggers a specific *FAU* is usually known, at least in retrospect. It may be, for example, a certain governmental decision, a change in international trade, a major event leading to changes in expectations (such as the attack on the US World Trade Center on September 11, 2001) etc. Accordingly, one might predict an imminent *FAU* (possibly not so rigidly defined) when a triggering event occurs. This does not exclude predictability of *FAUs* with a longer lead-time, as in the present study. On the contrary, it seems plausible that a *FAU* may be triggered only if and when the situation becomes "ripe" for a *FAU*; otherwise the government would not make that decision; unemployment would be less sensitive to international trade, etc. If that conjecture is correct, the approach under consideration can predict such a "ripe" situation. This may be of independent interest for the modeling and containment of unemployment, and it may help in the short-term prediction of *FAU*s.

(v) The dynamics of the economy, including changes in unemployment, may have some basic features that are common in various complex processes and found in models of statistical physics (Burridge and Knopoff, 1967; Gell-Mann, 1994; Newman et al., 1994; Allègre et al., 1995; Holland, 1995; Blanter et al., 1997; Gabrielov et al., 2000a; Keilis-Borok and Soloviev, 2003). The results presented in this section may provide for such modeling of the scaling and parameterization of the processes considered and a formal definition of the relevant critical phenomena.

The data. Databases issued by the Organization for Economic Cooperation and Development (OECD, 1997) and the International Monetary Fund (IMF, 1997) have been used. Past *FAU*s were identified by an analysis of the monthly statistics of unemployment. To explore the predictability of *FAU*s the monthly indicators listed below were analyzed. For France the data sources are sufficiently complete for the time period between January 1965 and May 1997, and it is this period that is considered here.

Composite characteristics of the French national economy:

1. *IP*: Industrial production index, composed of weighted production levels in numerous sectors of the economy, in % relative to the index for 1990.

2. *L*: Long-term interest rate on 10-year government bonds, in %.

3. *S*: Short-term interest rate on 3-month bills, in %.

The analogues of these indicators for the USA have been successfully used in research on predicting American economic recessions (Stock and Watson, 1993; Keilis-Borok et al., 2000).

Characteristics of more narrow areas of economy which are sensitive to its overall state:

4. *NC*: The number of new passenger car registrations, in thousands of units.

5. *EI*: The expected prospects for the national industrial sector.

6. *EP*: The expected prospects for selected manufacturers.

7. *EO*: The estimated current volume of orders.

The last three indicators are subjective; they distinguish the "good" from "bad" situations by polling 2,500 manufacturers, with the expectations weighted by the size of their businesses. They are available only for France.

Two indicators related to the American economy:

8. *FF/$*: Value of U.S. dollar in French francs.

9. *AR*: The state of the American economy: is it close to a recession or not? This indicator shows whether a pre-recession alarm for the USA is or is not declared by the algorithm described by Keilis-Borok et al. (2000). Note that it is not the American recessions themselves are considered here, but a certain precursor to these recessions, which includes no European indicators.

Identification of *FAU*s. Here, the definition of *FAU* that was qualitatively described above (Fig. 22) is formalized. As before $u(m)$ is the number of unemployed in a month m (m = 1, 2, ...). *FAU*s are defined as follows. First, smoothing out the seasonal variation of u function $U(m) = W^u(m/m-6, m+6)$ is obtained that is a value of the regression (1) over the time interval ($m - 6$, $m + 6$). Next, a function $F(m/s) = K^U(m+s, m) - K^U(m, m-s)$ is determined that is the difference between the linear trends in regression (1) of $U(m)$ within s subsequent months and s preceding months. This function with s = 24 months is used as a coarse measure of unemployment acceleration. Finally, the *FAU*s are defined by the local maxima of $F(m)$ exceeding a certain threshold **F**. The time m^* and the height F^* of such a maximum are, respectively, the time and the magnitude of a *FAU*. Acceleration ends in a month m_e of the subsequent local minimum of $F(m)$.

FAUs in France. Monthly unemployment in France is shown in Fig. 24. The function $F(m)$ for the time period considered, from January 1965 to May 1997, is also shown in Fig. 24. One may see in Fig. 24 that the theshold **F** = 4 identifies obviously outstanding peaks of $F(m)$. Seven *FAU*s identified by the condition $F^* \geq$ **F** = 4 are listed in Table 19. One can see that each *FAU* is the beginning of a long unemployment surge, lasting 16 – 24 months. Since it is determined after a strong smoothing of unemployment rate, the meaningful accuracy of prediction may hardly be better than about 2 months. Three "major" *FAU*s, marked in bold in Table 19, are distinctly larger than the others.

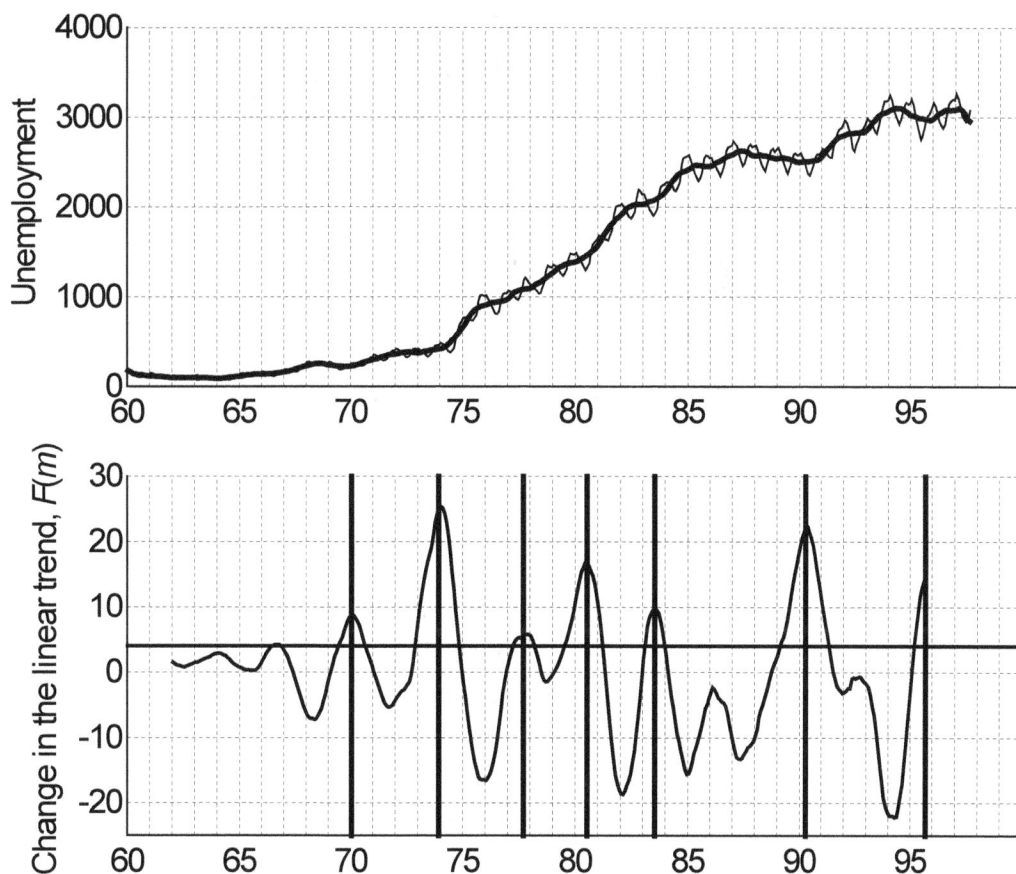

Figure 24. Unemployment in France. *Top:* Monthly unemployment, thousands of people. Thin line: $u(m)$, data from the OECD database; note the seasonal variations. Thick line: $U(m)$, data smoothed over one year. *Bottom:* Determination of FAUs. $F(m)$ shows the change in the linear trend of unemployment $U(m)$. FAUs are attributed to the local maxima of $F(m)$ exceeding threshold $F = 4.0$ shown by horizontal line. The thick vertical lines show moments of the FAUs.

Besides the magnitude F^*, the scale of FAU phenomenon may be characterized by the "*glut*" of unemployment G (Fig. 25)

$$G = \sum \{U(l) - W^U(l/m^* - s, m^*)\}, \qquad m^* \leq l \leq m_e. \qquad (4)$$

Here m^* and m_e are as defined above the first and the last month of the FAU. The first term is the actual cumulative loss of employment, in person-months, during the FAU. The second term shows what this loss would have been had unemployment not accelerated but rather grew at the same rate as right before the *FAU*.

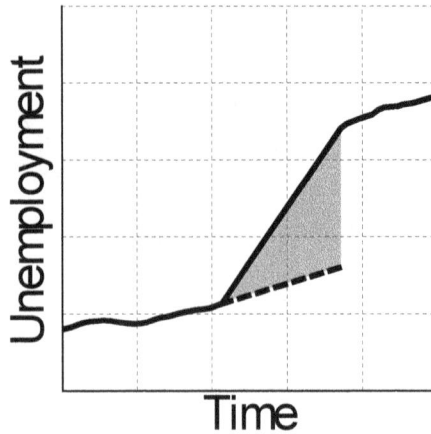

Figure 25. Glut of unemployment: schematic definition. Its measure is the shaded area, showing the amount of person-months of jobs lost. The dashed line is the extrapolation of the pre-*FAU* trend.

The values of *G* calculated for *FAU*s by means of (4) are given in the last line of Table 19.

Table 19. *FAU*s in France, 1965–1997

Time, *year: month*	1970:01	1974:01	1977:09	1980:07	1983:07	1990:05	1995:09
"Magnitude" F^*	7.6	22.6	5.3	15.7	9.2	20.3	9.4
Duration (m_e - m^*) (months)	22	24	16	20	19	21	20
G (person × months)	1993	5570	371	2760	1400	3420	2960

5.2 Hypothetical Prediction Algorithm

Premonitory trends of single indicators. Here, the "premonitory" trends of the indicators are explored, which tend to occur more frequently as a *FAU* approaches.

The trends are approximated by the regression coefficients $K^f(m/s)$ that are defined accordingly to formulas (1), (2) with f replaced by the symbol of an indicator. The value of $K^f(m/s)$ may be used for prediction, since it is attributed to the *end* of the time window where it is determined; accordingly, it does *not* depend on the information about the future. (The definition of the *FAUs* is applicable only in retrospect, two years after a *FAU* occurs in order to ensure a reliable identification of past *FAUs*).

An example of further analysis is given in Fig. 26. It shows the indicator $L(m)$ and its trend $K^L(m/s)$. The plot of $K^L(m/s)$ suggests that its peaks (i.e. the relatively steep upward trends of $L(m)$) emerge more frequently before the *FAUs*; in that sense they are "premonitory" to *FAUs*. To give a robust quantitative definition of a premonitory trend, the values of $K^L(m/s)$ are defined on the lowest level of resolution, distinguishing only the values above and below a threshold $T^L(Q)$, i.e. the discretization is made as described above in Section 4.2. The threshold is defined as a percentile of a level Q, that is, by the condition that $K^L(m/s)$ exceeds $T^L(Q)$ during $Q\%$ of the months considered.

Similar analyses of other indicators are summarized in Table 20. Premonitory behavior of these indicators has a transparent qualitative explanation.

(i) A steep rise in macroeconomic indicators (Nos. 1 – 3 in Table 20) reflects the "overheating" of the economy. For industrial production (No. 1) this may sound counterintuitive, since the rise of production is supposed to create more jobs. However, premonitory to *FAUs* is only a steep rise, above 50% percentile, which creates oversupply.

(ii) A steep decline of the next four indicators reflects the expectations of the general public (#4) and the appraisal of economic trends by the business community (#5 – 7). These indicators seem better substantiated, than, say, public opinion polls.

(iii) The last two indicators (#8 and 9) are related to the American economy. Their premonitory behavior would be more difficult to foresee – an equally plausible explanation could be probably found for the opposite behavior. Accordingly, these indicators yield particularly strong constraints on the modeling of the dynamics of unemployment.

Summing up, the description of the unemployment-relevant situation has been reduced to a monthly time series of a binary vector with 9 components, as is usual in the pattern recognition of infrequent events. For convenience, the same code, 1, is given to the "premonitory" trend of each indicator, regardless of whether it is an upward or a downward one.

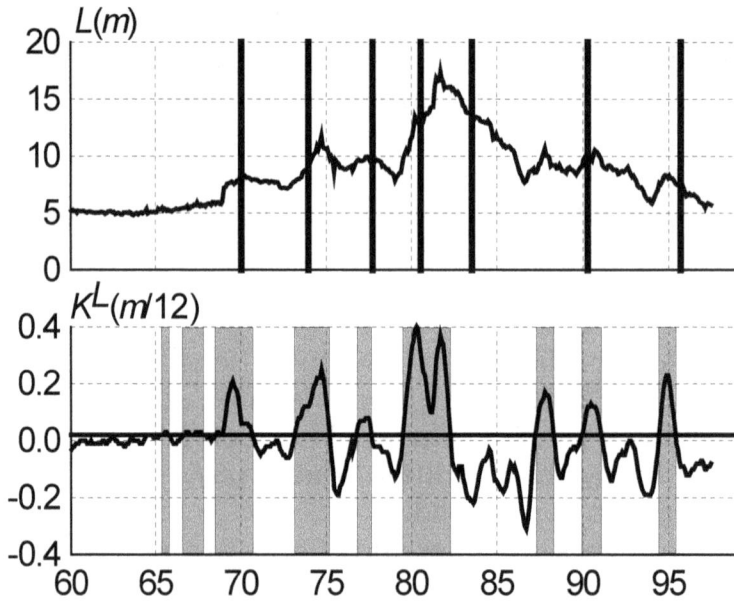

Figure 26. Robust discretization of an economic indicator. Top: $L(m)$, interest rate on long-term (10 year) governmental bonds (%). Source: OECD. Bottom: $K^L(m/12)$, linear growth of $L(m)$ in the sliding time window $(m - 12, m)$. The thin horizontal line is the upper 33% percentile for $K^L(m/12)$ so that it exceeds this threshold 33% of time. The gray bars mark time intervals when the value of $K^L(m/12)$ exceeds the threshold. The thick vertical lines show moments of *FAU*s. Note that after a quiet period before 1967 "large" values of $K^L(m/12)$ tend to appear more frequently when a *FAU* is approaching. However, with this threshold they would "predict" the *FAU*s only in combination with other indexes.

Ensemble of premonitory trends. Here, it is considered how the approach of a *FAU* is reflected in the *collective behavior* of the indicators. Its simplest description is function Δ(*m*) – the number of non-premonitory indicators at the month *m*. If the identification of premonitory trends is correct then the value of Δ(*m*) should decrease as a *FAU* approaches. By definition Δ(*m*) is the number of zeros in the binary code of the situation. This is the so-called "Hamming distance" between that code and the code of the "perfect" premonitory situation, when all the components are equal to 1, that is, all the trends are premonitory. This measure was used in different applications of the pattern recognition – to elections (Lichtman and Keilis-Borok, 1989), recessions (Keilis-Borok et al, 2000), and earthquakes (Keilis-Borok and Kossobokov, 1990; Vorobieva and Levshina, 1994; Keilis-Borok and Soloviev, 2003).

Table 20. Trends and thresholds

Indicator	Premonitory trend	s	Q (%)
IP: industrial production index	Upward	12	50
L: interest rate, long-term bonds	Upward	12	33
S: interest rate, short-term bills	Upward	12	25
NC: new passenger cars registrations	Downward	6	33
EI: prospects for industrial sector	Downward	6	33
EP: prospects for selected manufacturers	Downward	6	33
EO: orders	Downward	6	33
FF/$: french francs per USD, exchange rate	Downward	6	33
AR: recession alarm in the U.S.	Is current		

Only the composite indicators *IP*, *L*, and *S* were considered first. The value of Δ(*m*) may vary in this case from 0 to 3. Change of Δ(*m*) through the time considered is juxtaposed with *FAU*s in Fig. 27. One can see that the minimal value Δ(*m*) = 0 appears within 1 – 12 months before a *FAU* and at no other time. Data on Fig. 27 suggest the following hypothetical prediction algorithm: *An* *alarm is declared for 6 months after each month with* $\Delta(m) = 0$ *(regardless of whether this month belongs or not to an already determined alarm).* A waiting period of 6 months is introduced because in three cases (1977, 1980, and 1995) the premonitory pattern does not appear right before a *FAU*. Results of prediction are shown in the top row of Fig. 28. One can see, that this algorithm predicts 6 out of 7 *FAU*s, including all three major ones.

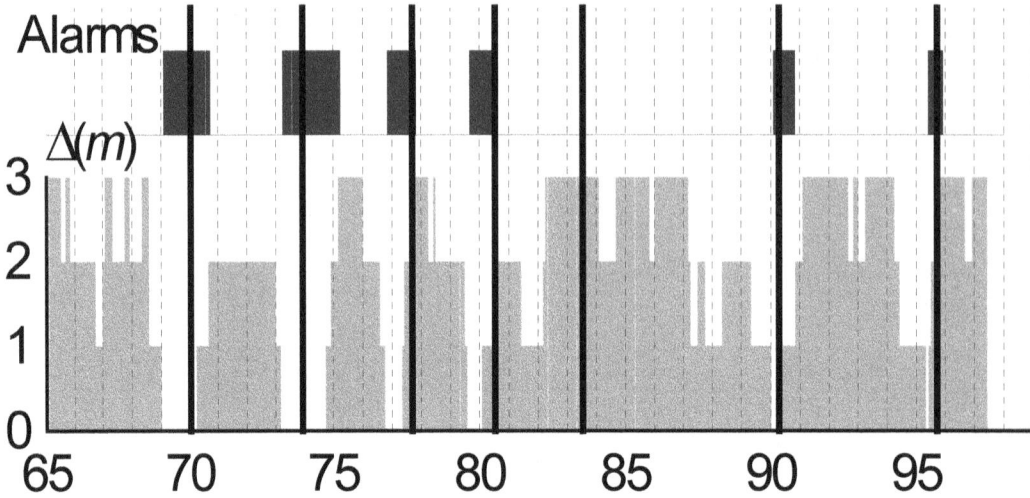

Figure 27. Collective performance of premonitory trends. Function $\Delta(m)$ is the number of *non-premonitory* trends at month m. Vertical lines show *FAU*s. Alarms (shown by dark gray bars) are declared for 6 months after each month when $\Delta(m) = 0$.

5.3. Stability of Prediction (Sensitivity Analysis)

The above algorithm is developed retrospectively, with a certain freedom in the ad hoc choice of indicators and numerical parameters. The results of prediction should be stable to their variations. To estimate this stability, the prediction needs to be repeated with different groups of indicators and values of parameters. In this study, the algorithm formulated above is generalized as follows: *An alarm is declared for r months after each month with* $\Delta(m) \leq D$ *(regardless of whether this month belongs or not to an already determined alarm).* The hypothetical prediction algorithm formulated above corresponds to the case $D = 0$, $r = 6$.

Five other groups of indicators (Table 21) were considered. The results are compared in the error diagram in Fig. 29. Error diagrams as a tool for the evaluation of a prediction method and for the optimal choice of a response to a prediction are described by Molchan (1997, 2003). Numerical values of τ (the total duration of alarms in per cent of the total time considered), n (the number of failures to predict), and f (the number of false alarms) are given in the last three columns of Table 21.

Table 21. Variation of the groups of indicators used for prediction

Indicators	D	r	τ (%)	n	f
IP, L, S	0	6	23	1	0
IP, L, S, FF/$	0	6	13	3	1
NC, EO	0	3	31	0	5
IP, L, S, NC, EO	1	6	26	1	1
IP, L, S, NC, EO, FF/$	1	6	19	2	1
IP, L, S, NC, EO, AR	2	6	28	1	0

The last three columns show the performance of a prediction algorithm.

One can see that composite indicators (group 1) give relatively the best predictions, alone or with some other indicators. Only the group 3 gives no failures to predict; however that is achieved at the price of 5 false alarms, which are hard to accept. There were attempts to replace the indicator *EO* in groups 3 – 6 by a subjective indicator *EI* or *EP*, but in each case the number of failures to predict and false alarms increased simultaneously.

Numerical parameters that define premonitory trends (Q^I, s), alarms (D, r) and *FAUs* (**F**, s) were also varied.

Stability of prediction (sensitivity analysis). The stability of these predictions has been explored by repeating them with various indicators and parameters and comparing the ensuing alarms. The percentile Q was varied from 5 to 10% and the window s for determination of the regression coefficients was varied from 9 to 15 months. The outcomes of several experiments on the variation of the adjustable parameters are shown in Table 22. These experiments were made for the prediction based on the first group of indicators: *IP*, *L*, and *S* (line 1 in Table 21).

Variation of definition of FAU. Two adjustable parameters are used to define the moments of *FAU*s: the time window *s* for evaluation of the trend of unemployment and the threshold **F** for identification of "fast" acceleration. The analysis given above corresponds to the case *s* = 24 months and **F** = 4. Variation of *s* from 21 to 27 does not shift the moments of *FAU*s by more than 1-2 months – quite within the meaningful accuracy of prediction. The variation of **F** in the range from 3.6 to 5.0 does not lead to the emergence or disappearance of the targets of the prediction – the local maxima of *F*(*m*).

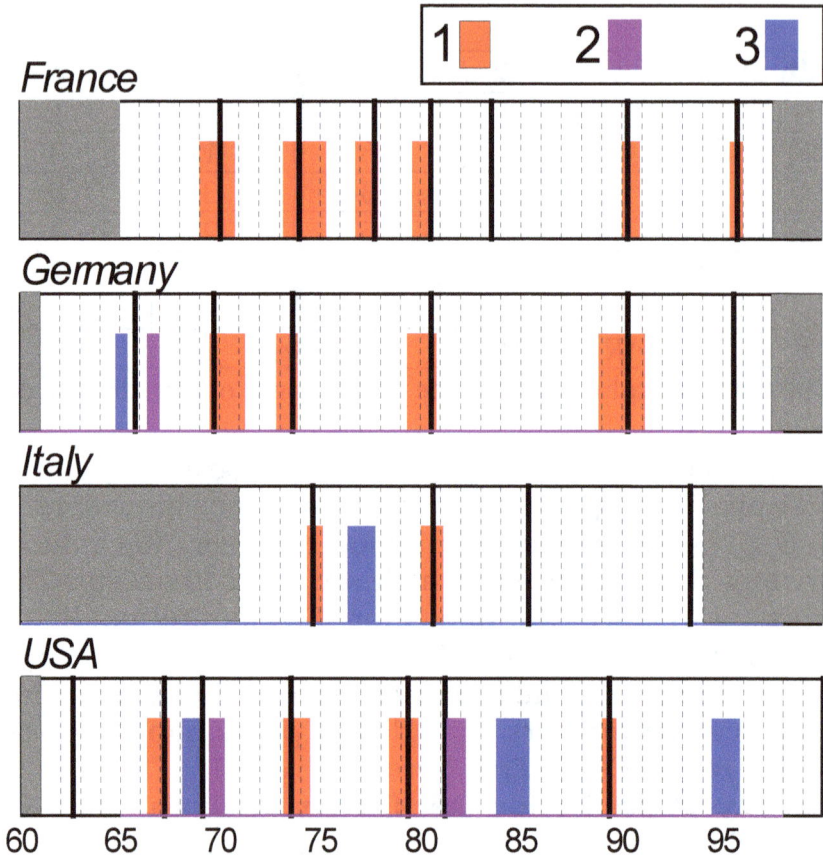

Figure 28. **Retrospective prediction for four countries.** *FAU*s and alarms obtained by the prediction algorithm. The thick vertical lines show the moments of *FAU*s in a country. Bars – the alarms with different outcome: 1 –

alarms that predict *FAUs*, 2 – alarms starting shortly after *FAUs* within the periods of unemployment surge, 3 – false alarms. Shaded areas on both sides indicate the times, for which data on economic indicators were unavailable. In case 2 *FAUs* are counted, for rigor sake, as not predicted. Note however that the alarms starting shortly after these *FAUs* are practically useful (see Section 5.4)

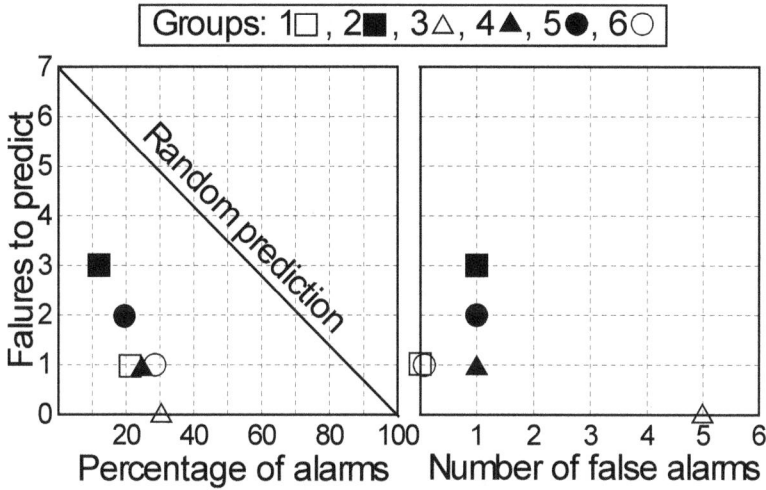

Figure 29. Error diagram for predictions with various groups of indicators. Group 1 includes *IP, L, S*; 2 - *IP, L, S, FF/$*; 3 - *NC, EO*; 4 - *IP, L, S, NC, EO*; 5 - *IP, L, S, NC, EO, FF/$*; 6 - *IP, L, S, NC, EO, AR*.

Summing up, one can conclude that the *most stable results are obtained with group 1 indicators - the three composite economic indicators*. Prediction with this group produced no false alarms; it sustains best the change of numerical parameters; and it does not gain by the addition of other indicators.

Next, the algorithm is tested on the independent data not used in its development. It is applied to Germany and Italy and to the USA.

Table 22. Variation of adjustable parameters for indicators *IP*, *L*, and *S*

#	Indicator(s), for which variation of a parameter is made	Variation of the parameter	τ (%)	n	f
Variation of Q					
1	IP	60%	23	1	0
2		40%	22	1	0
3	L	40%	28	1	1
4		25%	15	2	0
5	S	30%	30	1	2
6		20%	23	1	0
Variation of s					
7	IP, L, S	15	20	1	0
8		9	24	2	1

The last three columns summarize the performance of a prediction algorithm

5.4. Retrospective Application of the Algorithm for Several Countries

Germany and Italy. Here the algorithm formulated in Section 5.2 is applied to the data for Germany and Italy. *Exactly* the same algorithm is used, including the definition of the trends $K^f(m, m\text{-}s)$; the values of adjustable numerical parameters; and the rule for declaration of alarms. Such transfer from country to country is possible, since the algorithm is self-adaptive, where the percentiles Q^f common in each application determine the thresholds for discretization of the trends.

The results are shown in Fig. 28. Table 23 and Fig. 30 that compare the quality of prediction for different countries. Alarms preceding *FAU*s are usually continuing within the periods of fast unemployment growth (after *FAU*s), and such continuations should obviously not be regarded as false alarms. Alarms within the periods of fast unemployment growth are also not regarded as false ones. For Germany the quality of prediction is about the same as for France, while for Italy it is lower, though still better than random.

Table 23. Comparative performance of the prediction algorithm for France, Germany, Italy, and the USA

Country	FAUs			Alarms "not predicting" FAUs	
	Total	"Predicted"	"Missed"	After FAUs	"False"
France, 1965–97	7	6	1	none	none
Germany, 1961–97	6	4	2	1	1
Italy, 1971–93	4	2	2	none	1
USA, 1961-99	7	4	3	2	3

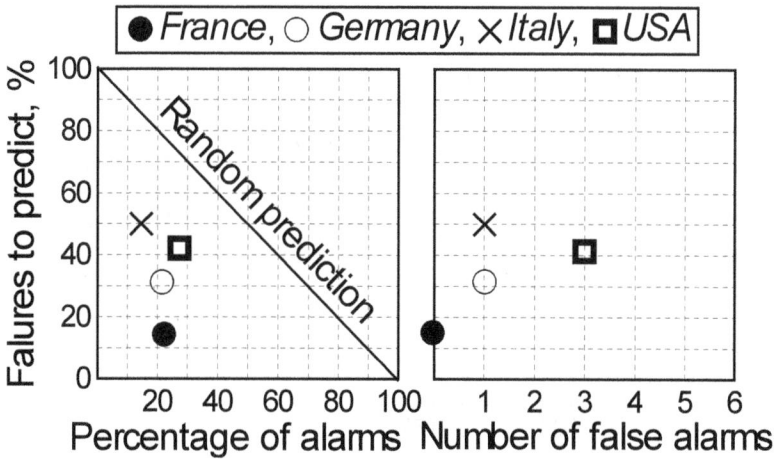

Figure 30. Error diagram for predictions for different countries.

Unemployment in the U.S. The data on monthly unemployment rates for the U.S. civilian labor force are used, as given in USDL (1999). Unlike Europe, unemployment in USA had no general trend during the years considered. One can see this in Fig. 31. The FAUs are the times when unemployment started to surge, that are the local minima of the unemployment rate. They are formally defined as follows. Let $R(m)$ be the smoothed monthly unemployment rate in a month m. Then $R(m)$ has the local minima in a month m^* if for $j = 1, 2, 3, 4$ $R(m^*-j) \geq R(m^*)$ and $R(m^*+j) > R(m^*)$. Seven such minima are identified within the period 1960-1999 in 1962:08 (9), 1967:03 (3), 1969:02 (28), 1973:07 (24), 1979:05 (19), 1981:03 (21), and 1989:05 (38). The duration of the unemployment surge is given in brackets after the corresponding months m^*, which are the targets of the prediction.

Application of the algorithm to the U.S. Indicators *IP*, *S*, and *L* have the following American equivalents. For *IP* – "industrial production, total". This is the index of the real (constant dollars) output in the entire economy (dimensionless). For *S* - interest rate on 90-day U.S. treasury bills, at an annual rate (in percent). For *L* - interest rate on 10-year U.S. treasury bonds, at an annual rate (in percent). Time series of these indicators were used for prediction of economic recessions in the U.S. (Section 4.2) and obtained from the CITIBASE, where *IP*, *FYGM3*, and *FYGT10* are their respective mnemonics.

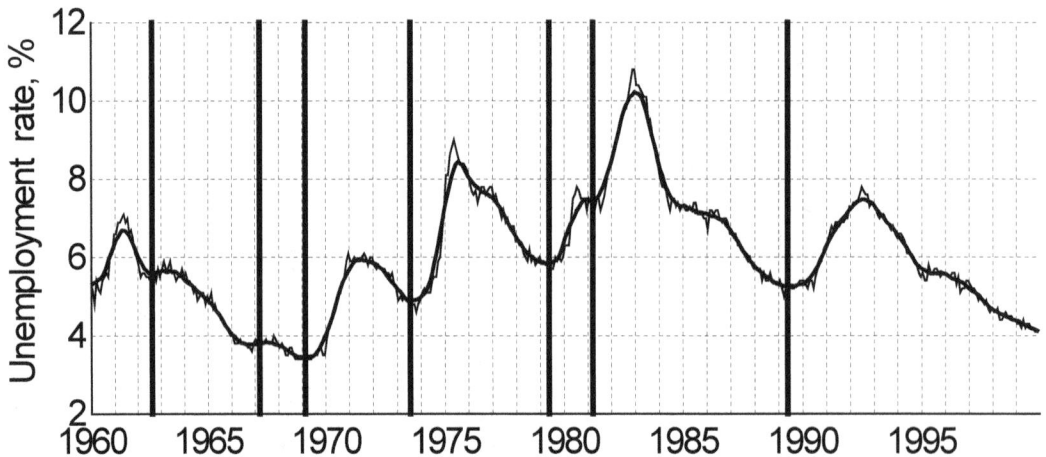

Figure 31. **Unemployment rates in the U.S.** Thin line: $r(m)$, original data. Thick line: $R(m)$, data after smoothing out the seasonal variations. The thick vertical lines show the moments when unemployment started to surge (local minima of smoothed unemployment rate).

Alarms and *FAU*s are juxtaposed in the bottom row of Fig. 28. One can see that 4 out of 7 *FAU*s are captured by alarms; three *FAU*s, in 1962, 1969, and 1981, are missed; and there are false alarms, in 1968, 1983, and 1994. The alarms within the periods of unemployment growth are not regarded as false ones. The total count of errors for the *USA* is given in Table 23 and Fig.24. It is worse than for European countries, though still better than random.

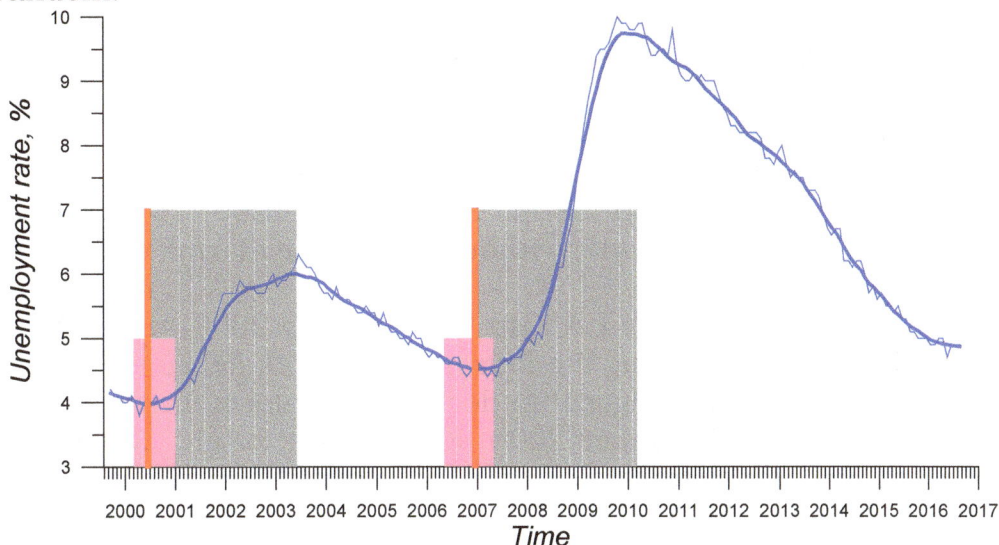

Figure 32. Prediction of future *FAU*s in the U.S. from September 1999. Thin blue curve shows monthly unemployment rate in the U.S., according to the data of the Bureau of Labor Statistics, U.S. Department of Labor. Thick curve shows this rate with seasonal variation smoothed away. Vertical red lines show prediction targets – the moments of *FAU*; grey bars – the periods of unemployment's growth; pink bars – periods of alarms.

Note, that this is a rigorous count, giving lower estimate for the algorithm's performance. Such an estimate is necessary for some purposes, e.g. evaluation of statistical significance of predictions, but for other purposes it might be misleading. Next a more practical estimate is discussed. Let us consider the count of errors from a disaster preparedness point of view. One of the alarms ended in 1968:12, a month before the *FAU*; it is counted as a false alarm and the subsequent *FAU* – as missed by prediction. Similarly, the *FAU* in 1981:03 is counted as missed, while it was followed by an alarm a month later. Since a *FAU* is a starting point of a long surge in unemployment, lasting about 20 months, a one month difference is not necessarily important for a decision-maker responding to a prediction. Moreover, this difference is within the accuracy of *FAU*s, since they are determined after considerable smoothing of the unemployment rate. Accordingly, for the end user only the three errors might be worth counting: the failure to predict in 1962 and the false alarms in 1983 and 1994.

5.5 Application to the U.S. Advance Prediction

Prediction of the future *FAU*s was launched for USA by the algorithm formulated in Section 5.2. The results are presented in Fig. 32. It shows that from September 1999 two correct predictions have been made, without ether false alarms or failures to predict.

Analyzing the data for USA up to December 2000, it has been found that $\Delta(m) = 0$ during the four months, from February to May 2000. Accordingly, the algorithm declared the alarm for the period from February to November 2000 (Keilis-Borok et al., 2005). The data on unemployment rates for the subsequent period, up to July 2002, confirmed that prediction of a *FAU* materialized in July 2000.

The next alarm was declared after April 2006 when $\Delta(m) = 0$ (Keilis-Borok et al., 2006) for the first time after 2000 during the seven months periof from April to October of 2006. In November 2006 this second prediction was filed on the web site of the Anderson School of Management, University of California, Los Angeles (http://www.uclaforecast.com). The relevant FAU occurred in December 2006.

Obviously, these confirmations are most encouraging, but much longer experimentation is necessary to validate the algorithm.

Numerous other warnings of a coming surge of unemployment in the USA did appear in 2000 and 2006, and even in the popular media. The particular feature of the prediction discussed here, however, is that it was obtained using a formal unambiguous algorithm, and that it indicates a specific time interval when the unemployment will start to surge.

5.6 Recessions and Unemployment

Table 24 shows the comparison of the periods of unemployment growth and recession in the U.S. We see that all eight American recessions during the time period, 1960-2009, did occur within the eight longest periods of unemployment growth. Three of the seven pre-recession alarms determined by the algorithm developed Keilis-Borok et al. (2000) and described in Section 4.3 begin before the relevant months m^* (Table 24). The level of unemployment, measured in a different way, was actually used for the determination of these alarms. This confirms that the prediction of unemployment is indeed not equivalent to the prediction of recessions.

Table 24. Periods of unemployment growth, recessions in the U.S., and pre-recession alarms

Local minima m^*	End of unemployment growth	Recessions	Pre-recession alarms
1959:11	1961:05	1960:05 – 1961:02	
1962:08	1963:05		
1967:03	1967:06		
1969:02	1971:06	1970:01 – 1970:11	1969:07 – 1969:12
1973:07	1975:07	1973:12 – 1975:03	1973:06 – 1973:11
1979:05	1980:12	1980:02 – 1980:07	1979:03 – 1980:01
1981:03	1982:12	1981:08 – 1982:11	1981:02 – 1981:07
1989:05	1992:07	1990:08 – 1991:03	1989:06 – 1990:07
2000:06	2003:05	2001:04 – 2001:11	2001:01 – 2001:03
2006:12	2010:02	2008:01 – 2009:06	2008:05 – 2008:09

CHAPTER 6

Prediction of Homicide Surges in Megacities

Dynamics of crimes reflects important aspects of sustainability of our society and the risk of its destabilization – a prelude to a disaster. Keilis-Borok et al. (2003) have considered a prominent feature of crime dynamics – surge of the homicides in a megacity. This study integrates the professional expertise of the police officers and of the scientists working on pattern recognition of infrequent events. The latter is a type of artificial intelligence methodology that has been successful in predicting infrequently occurring phenomena that result from highly complex processes.

Statistics of several types of crimes in the city of Los Angeles over the period 1975-2002 has been analyzed by Keilis-Borok et al. (2003). This analysis focuses on how these statistics change before a sharp and lasting rise ("a surge") of the homicide rate. The goal is to find an algorithm for predicting such a surge by monitoring the rates of different crimes.

The hope for feasibility of this goal comes from two sources. The first one is the set of available crime statistics, showing that a surge of major crimes is preceded by the rise of less severe crimes. The second one is recent research in the predictions of critical phenomena (i.e. abrupt overall changes) in various complex non-linear systems, such as those in theoretical physics, earth sciences, social sciences, etc.

Data. Out of a multitude of relevant data statistics of robberies, assaults, burglaries, and the homicides themselves is analyzed.

Results. The findings may be summarized as follows: Episodes of a rise of burglaries and assaults simultaneously occur 4 to 11 months before a homicide surge, while robberies decline. Later on, closer to the rise in homicides, robberies start to rise. These changes are given unambiguous and quantitative definitions, which are used to formulate a hypothetical algorithm for the prediction of homicide surges.

In retrospective analysis it has been found that this algorithm is applicable through all the years considered, despite substantial changes both in socio-economic conditions and in the number of crimes. Moreover, it gives satisfactory results for the prediction of homicide surges in New York City as well. Sensitivity tests show that predictions are stable to variations of the adjustable elements of the algorithm.

What was learned? The existing qualitative portrayals of crime escalation are complemented here by a quantitatively defined set of precursors to homicide surges. The same set emerges before each surge through the time period under consideration. This implies the existence of a "universal" scenario of crime escalation, independent of concrete reasons triggering each surge. These findings provide heuristic constraints for the modeling of crime dynamics and indicate promising lines of further research.

Perspective. Decisive validation of these findings requires experimentation in *advance prediction*, for which this study sets up a base. Particularly encouraging for this further research is the wealth of yet untapped possibilities: so far only a small part of the data and mathematical models that are currently available and that are relevant to crime dynamics have been used.

On the practical side, these results enhance the capability to identify a situation that is "ripe" for homicide surges and, accordingly, to escalate the crime prevention measures. In a broader scheme of things, a surge of crime is one of potential ripple effects of natural disasters. Accordingly, the risk of a natural disaster is higher in such a situation.

Prediction of a specific phenomenon in crime dynamics: a large and lasting increase in the homicide rate was considered. Qualitatively, this phenomenon is illustrated in Fig. 33; Keilis-Borok et al. (2003) called it by the acronym **SHS, for "Start of the Homicide Surge"**. The goal is to find a method to predict an *SHS* by monitoring the relevant observed indicators, i.e. to recognize the "premonitory" patterns formed by such indicators when an *SHS* approaches. In terms of pattern recognition, an algorithm (a "recognition rule") that solves the following problem is looked for:

given the time series of certain crime rates (or of other relevant indicators) prior to a moment of time *t*,

to predict whether an episode of *SHS* will or will not occur during the subsequent time period $(t, t+\tau)$; in other words, whether the lasting surge of homicides will or will not start during that period.

If the prediction is "yes", this period will be the "period of alarm." The possible outcomes of such a prediction are illustrated in Fig. 34.

The probabilistic component of this prediction is represented by the estimated probabilities of errors – both false alarms on one side and failures to predict on the other. That probabilistic component is inevitable since we consider a highly complex non-stationary process using imprecise crime statistics. Moreover, the predictability of a chaotic system is, in principle, limited.

Such "yes or no" prediction of specific extraordinary phenomena is different from predictions in a more traditional sense - extrapolation of a process in time, which is better supported by classical theory.

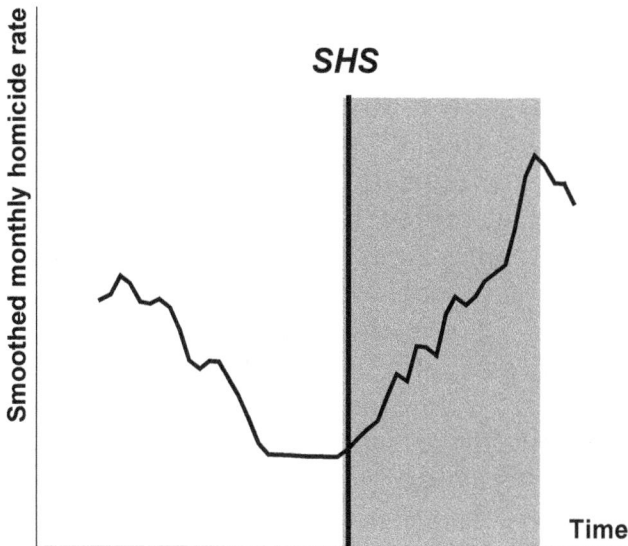

Figure 33. Target of prediction – the Start of the Homicide Surge ("*SHR*"); schematic definition. Gray bar marks the period of homicide surge.

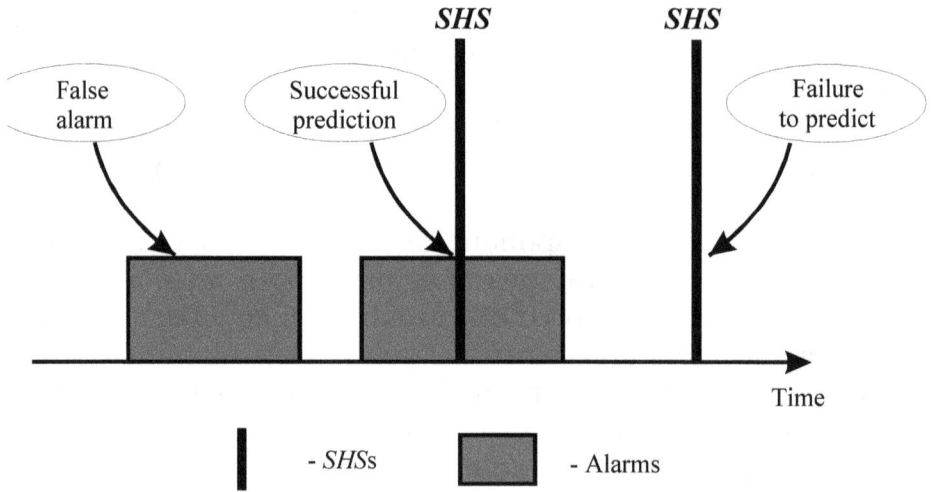

Figure 34 Possible outcomes of prediction.

The methodology used by Keilis-Borok et al. (2003) is *pattern recognition of infrequent events* that was developed by the artificial intelligence school of the mathematician I. M. Gelfand (Gelfand et al., 1976) for the analysis of infrequent phenomena of highly complex origin. Using this methodology, a so-called "technical" analysis that involves a heuristic search for phenomena preceding episodes of *SHS* is conducted here. A distinctive feature of this methodology is the robustness of the analysis, which helps to overcome both the complexity of the process considered and the chronic imperfection of the data; in that aspect it is akin to exploratory data analysis, as developed by the statistics school of J. Tukey (Tukey, 1977). Robust analysis – "a clear look at the whole" – is imperative in a study of any complex system (Gell-Mann, 1994). The surest way *not* to predict such a system is to consider it in too fine detail (Crutchfield et al., 1986).

Pattern recognition of infrequent events has been successfully used in geophysics, geological prospecting, medicine, and many other areas. Close to the present study are recent studies of the prediction of economic recessions and surges of unemployment (Keilis-Borok et al., 2000, 2005). The same pattern recognition algorithm, called "Hamming distance," is used by Keilis-Borok et al. (2003). It has been applied in these studies (Keilis-Borok et al., 2000, 2005), as well as in predictions of American elections (Keilis-Borok and Lichtman, 1993) and in seismology (e.g. Kosobokov, 1983; Vorobieva and Levshina, 1994; Vorobieva, 1999). The essence of the algorithm will be clear from the way of crime statistics analyzing.

Following is a schematic outline of our analysis.

Data comprise the monthly rates of homicides, robberies, assaults, and burglaries for Los Angeles, 1975 – 2002 (Section 6.1).

Five targets of prediction (*SHS*) are defined during the time period under consideration (Sections 6.2, 6.6). Those are the moments when a years-long trend of the homicide rate turns from a decline to a long steep rise.

Premonitory changes of crime statistics have been found as illustrated in Fig. 35. Within several months before a homicide surge, burglaries and assaults simultaneously escalate, while robberies decline (Section 6.3). On the basis of these changes a *hypothetical prediction algorithm* is suggested (Section 6.4). In retrospect, it provides a robust satisfactory prediction (Sections 6.5, 6.6). However it has to be further validated by application to independent data. As always in prediction research, *the final validation of the algorithm requires prediction in advance*, for which this study sets up a base.

Later on, closer to a homicide surge, robberies also escalate (Section 6.7). These changes will be explored elsewhere.

Common notation. The analysis focuses on trends in the crime rates. These trends are estimated by linear regression, using the following notations.

$$C(m), \ m = 1,2...,$$

is the time series of a monthly indicator, where m is the sequence number of a month.

$$W^C(m/q, p) = K^C(q, p)m + B^C(q, p), \ q \le m \le p, \tag{5}$$

is the local linear least-squares regression of the function $C(m)$ within the sliding time window over the time period (q, p).

6.1. The Data

The following data sources are used:

(i) The National Archive of Criminal Justice Data (NACJD), placed on the web site:

http://www.icpsr.umich.edu/NACJD/index.html
(see Carlson (1998) for its description. This site contains data
for the years 1975-1993).

(ii) Data bank of the Los Angeles Police Department (LAPD
Information Technology Division); it contains similar data for the
years 1990 – May 2001.

Out of numerous crime statistics given in these sources, the
monthly rates of the four types of crime listed in Table 25 are
analyzed: homicides, robberies, assaults, and burglaries.

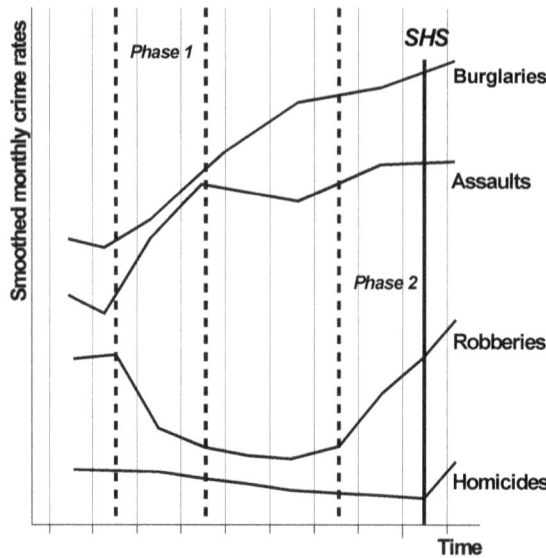

Figure 35. Scheme of premonitory changes in crime statistics.

6.2 Prediction Targets

Here and in the next two sections the data for 1975 – 1993 as
taken from the National Archive of Criminal Justice Data (*Carlson,
1998*) are analyzed.

Definition. Let $H(m)$, $m = 1,2...$, be the time series of the
monthly number of all homicides. Fig. 36 shows the plot of $H(m)$
in Los Angeles, per 3,000,000 inhabitants of the city. To identify
the episodes of *SHS* (Fig. 33) the seasonal variations, which are
clearly seen in Fig. 36, are smoothed out by replacing $H(m)$ by its
linear least square regression (5): $H^*(m) = W^H(m/m\text{-}6, m+6)$. Since
$H^*(m)$ is defined on the time interval $(m - 6, m + 6)$, it depends on

the future. Thus, it is admissible to define prediction targets (but not precursors).

Table 25. Types of crime considered
(after Carlson (1998); abbreviations are indicated in brackets)

Homicide	Robberies	Assaults	Burglaries
• All (H)	• All (Rob) • With firearms (FRob) • With knife or cutting instrument (KCIR) • With other dangerous weapon (ODWR) • Strong-arm robberies (SAR)*	• All (A)* • With firearms (FA) • With knife or cutting instrument (KCIA) • With other dangerous weapon (ODWA)* • Aggravated injury assaults (AIA)*	• Unlawful not forcible entry (UNFE) • Attempted forcible entry (AFE)*

* Analyzed in sensitivity tests only (Section 6.5)

6.3 Premonitory Trends of Single Types of Crime

Here the monthly data on seven types of crimes out of the 13 types listed in Table 25 are analyzed. "Premonitory" trends of each crime that tend to appear more frequently as an *SHS* approaches are looked for. Prediction itself is based on the collective behavior of these trends, as analyzed in the next Section. Orientation on a *set* of precursors has been found to be rather successful in prediction research: an ensemble of "imprecise" precursors usually gives better predictions than a single "precise" precursor (Keilis-Borok and Rotwain, 1990; Zaliapin et al., 2003b).

Observation. According to police experience, the crimes considered here often rise before an *SHS*.

To smooth out seasonal variations, the plot $C(m)$ of each type of crime is replaced by its regression (1): $C^*(m) = W^C(m/m\text{-}12, m)$. Regression is done over the prior 12 months and does not depend on the future, so that it can be used for prediction. These plots exhibit two consecutive patterns:

(i) First, one can see a simultaneous escalation of burglaries and assaults within several (4 to 11) months before an *SHS*; at the same time robberies are declining.

(ii) Later on, closer to an *SHS*, one can see, albeit not so clearly, a simultaneous escalation of different kinds of robberies.

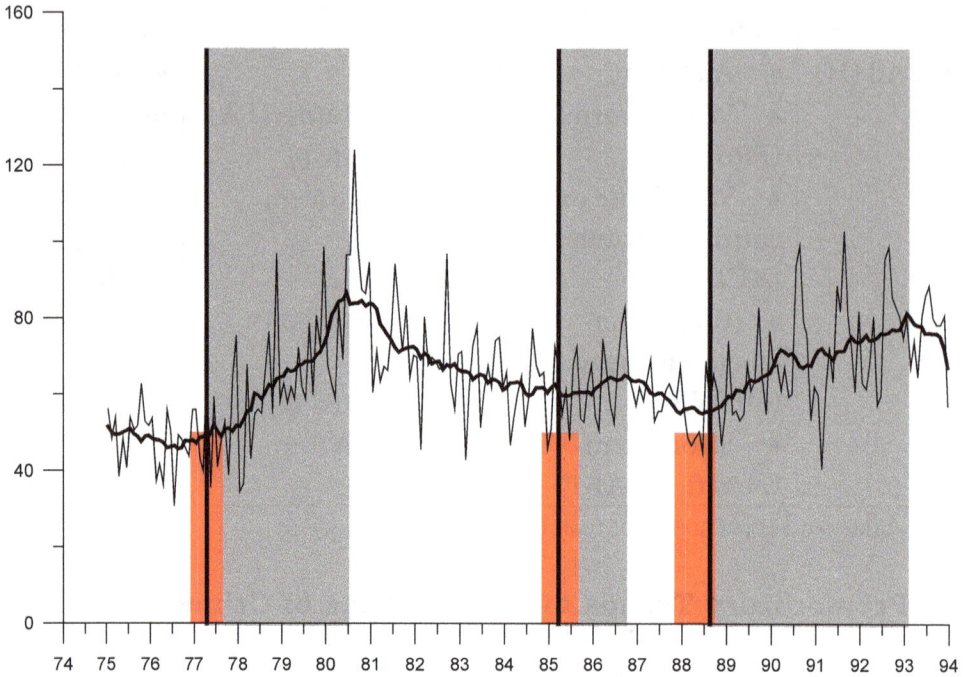

Figure 36. Total monthly number of homicides in Los Angeles city, 1975-1993. Data are taken from the National Archive of Criminal Justice Data (Carlson, 1998). Thin curve – original time series, *H(m)*, per 3,000,000 inhabitants. Thick curve - smoothed series *H*(m)*, with seasonal variations eliminated as described in Section 1. Vertical lines show the targets of prediction – episodes of *SHS* (Section 6.2). Gray bars are the periods of homicide surge. Red bars are the alarms declared by the hypothetical prediction algorithm (Section 6.4).

The first pattern is formally defined and explored in this study. The second pattern, briefly discussed in Section 6.7, will be explored elsewhere.

Discretization of crime trends. To quantify the above observation the trends of the crimes are approximated by the regression coefficients $K^C(m\text{-}s, m)$ where *C* identifies the type of crime. The value of K^C is attributed to the month m so that it does not depend on information on future months; therefore it can be used for prediction.

Next, following the pattern recognition approach, the trends (the values of K^C) are discretized on the lowest level of resolution: a binary one distinguishes only the trends above and below a threshold $T^C(Q^C)$. It is defined as a percentile of a level Q^C, that is, by the condition that $K^C(m\text{-}s, m)$ exceeds $T^C(Q^C)$ during Q^C percent of the months considered.

According to the above observations, it is expected that "premonitory" trends lay above the respective thresholds for assaults and burglaries, while they lay below these thresholds for robberies. One can see this in Fig. 37, showing the functions $K^C(m\text{-}12, m)$ for 7 crime types. For convenience, the same code, "1", will be given to the "premonitory" trend of each crime, regardless of whether it is above or below the threshold of discretization. The seven monthly crime statistics considered here are thus reduced to a binary vector with 7 components.

The crime statistics is discretized using the values of Q^C indicated in Table 26. The crime history, thus transformed, is given in Table 27.

Table 26. Premonitory trends for selected crime types

Crime type	Premonitory trend $K^C(m\text{-}s, m)$	s	Q^C, %	$T^C(Q^C)$
Rob	Below threshold	12	66.7	-3.69
FRob	"	12	66.7	-1.29
KCIR	"	12	50.0	1.73
ODWR	"	12	87.5	-3.87
FA	Above threshold	12	50.0	1.89
KCIA	"	12	50.0	-1.32
UNFE	"	12	50.0	1.94

See notations in the text.

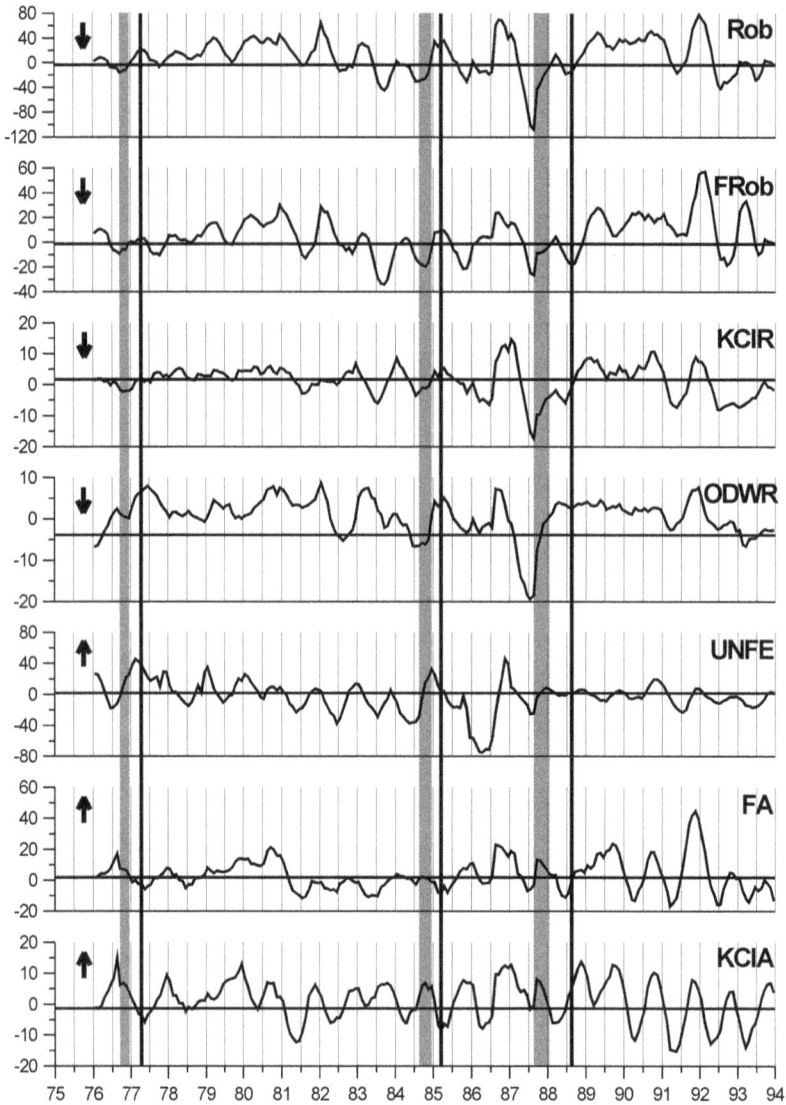

Figure 37. The regression coefficients $K^C(m\text{-}12, m)$ for seven crime types. See the definition in Section 6.3 and notations in Table. 25. Original data are taken from the National Archive of Criminal Justice Data (Carlson, 1998). Horizontal lines and arrows show respectively discretization thresholds and premonitory trends in accordance with Table 26. Vertical lines show episodes of *SHS*. Gray bars indicate months when $\Delta(m) \leq 1$.

Table 27. Binary codes of the trends for 7 types of crime and values of $K^H(m, m\text{-}12)$. See notations in Table 25. Discretization is defined in Table 26.

#	Year: Month	Rob	FRob	KCIR	ODWR	FA	KCIA	UNFE	Δ		K^H
1	1976: 01	1	0	0	1	0	1	0	4		0.44
2	1976: 02	1	0	0	0	1	1	0	4		0.21
3	1976: 03	1	0	0	0	1	1	0	4		-0.07
4	1976: 04	0	0	0	1	1	1	1	3		-0.28
5	1976: 05	0	0	0	1	1	0	1	4		-0.35
6	1976: 06	0	1	1	1	1	0	1	2		-0.33
7	1976: 07	0	1	1	1	1	0	1	2		-1.18
8	1976: 08	0	1	1	1	1	0	1	2		-0.96
9	1976: 09	0	1	1	1	1	0	1	2		-0.87
10	1976: 10	0	1	1	1	1	1	1	1	+	-0.79
11	1976: 11	0	1	1	1	1	1	1	1	+	-0.37
12	1976: 12	0	0	1	1	1	1	1	2		0.16
13	1977: 01	0	0	0	1	0	1	0	5		0.68
14	1977: 02	0	0	0	1	0	1	0	5		0.84
15	1977: 03	0	0	0	1	0	1	0	5		0.38
16	1977: 04	0	0	0	0	0	1	0	6		0.85
17	1977: 05	0	0	0	1	0	1	0	5		0.07
18	1977: 06	0	0	0	1	0	1	0	5		0.71
19	1977: 07	0	1	0	1	0	1	0	4		0.58
20	1977: 08	0	1	0	0	0	1	0	5		0.03
21	1977: 09	0	1	0	0	0	1	0	5		0.18
22	1977: 10	0	1	1	0	1	1	1	2		-0.17
23	1977: 11	0	1	0	0	1	1	1	3		0.24
24	1977: 12	0	1	0	0	1	1	1	4		0.81
25	1978: 01	0	0	0	0	1	1	1	4		0.38
26	1978: 02	0	0	0	0	1	1	1	4		0.16
27	1978: 03	0	0	0	0	1	1	1	4		0.53
28	1978: 04	0	0	0	0	0	0	0	7		-0.09
29	1978: 05	0	0	0	0	0	0	0	7		0.55
30	1978: 06	0	0	0	0	0	0	0	7		0.20
31	1978: 07	0	0	0	0	0	0	0	7		0.57
32	1978: 08	0	0	0	1	0	0	0	6		0.66
33	1978: 09	0	0	0	1	0	1	0	5		1.19
34	1978: 10	0	0	0	0	0	1	0	6		1.07
35	1978: 11	0	0	0	0	1	1	0	5		1.72
36	1978: 12	0	0	0	0	1	1	0	5		1.81
37	1979: 01	0	0	0	0	1	1	1	4		2.52
38	1979: 02	0	0	0	0	1	1	0	5		1.66
39	1979: 03	0	0	0	0	1	1	0	5		0.88
40	1979: 04	0	0	0	0	1	1	1	4		0.98
41	1979: 05	0	0	0	0	1	0	1	5		0.66
42	1979: 06	0	0	0	0	1	0	1	5		0.27
43	1979: 07	0	0	0	0	1	0	1	5		-0.21
44	1979: 08	0	1	0	0	1	0	1	4		-0.14
45	1979: 09	0	1	0	1	1	1	1	2		-0.36
46	1979: 10	0	0	0	0	1	1	1	4		0.45
47	1979: 11	0	0	0	0	1	1	1	4		0.19
48	1979: 12	0	0	0	0	1	1	1	4		2.15
49	1980: 01	0	0	0	0	1	1	1	4		1.82
50	1980: 02	0	0	0	0	1	1	1	4		1.49
51	1980: 03	0	0	0	0	1	1	0	5		0.82
52	1980: 04	0	0	0	0	1	1	0	5		1.08
53	1980: 05	0	0	0	0	1	1	0	5		0.63
54	1980: 06	0	0	0	0	1	0	0	6		1.51
55	1980: 07	0	0	0	0	1	0	1	5		1.86
56	1980: 08	0	0	0	0	1	1	1	4		2.79
57	1980: 09	0	0	0	0	1	1	1	4		3.26
58	1980: 10	0	0	0	0	1	1	1	4		2.56
59	1980: 11	0	0	0	0	1	1	1	4		2.43
60	1980: 12	0	0	0	0	1	1	1	4		2.22
61	1981: 01	0	0	0	0	1	1	0	5		1.78
62	1981: 02	0	0	0	0	1	1	0	5		0.85
63	1981: 03	0	0	0	0	0	0	0	7		-0.57
64	1981: 04	0	0	0	0	0	0	0	7		-2.02
65	1981: 05	0	0	0	1	0	0	0	6		-2.64
66	1981: 06	0	1	0	1	0	0	0	5		-3.35
67	1981: 07	0	1	1	1	0	0	0	4		-2.51
68	1981: 08	0	1	1	1	0	0	0	4		-2.10
69	1981: 09	0	1	0	1	0	1	1	3		-0.92
70	1981: 10	0	1	0	1	0	1	1	3		-0.02
71	1981: 11	0	0	0	1	0	1	1	4		-0.10
72	1981: 12	0	0	0	1	0	1	1	4		0.09
73	1982: 01	0	0	0	1	0	1	1	4		0.70
74	1982: 02	0	0	0	1	0	0	0	6		-0.63
75	1982: 03	0	0	0	1	0	0	0	6		-0.27
76	1982: 04	0	0	0	1	0	0	0	6		-0.71
77	1982: 05	0	0	0	1	0	0	0	6		-0.97
78	1982: 06	0	0	1	1	0	0	0	5		-1.31
79	1982: 07	1	1	1	1	0	0	0	3		-1.32
80	1982: 08	1	1	1	0	0	0	0	4		-0.63
81	1982: 09	1	1	1	0	0	0	0	4		0.73
82	1982: 10	0	1	0	0	0	1	1	4		0.45
83	1982: 11	0	1	1	0	0	1	1	3		0.43
84	1982: 12	0	1	0	0	0	1	1	4		0.33
85	1983: 01	0	0	0	0	0	1	1	5		0.48
86	1983: 02	0	0	0	0	0	1	0	6		-0.35
87	1983: 03	0	0	0	1	0	0	0	6		-1.32
88	1983: 04	0	0	0	1	0	0	0	6		-0.62
89	1983: 05	0	1	0	1	0	0	1	4		-0.22
90	1983: 06	0	1	1	1	0	0	1	3		-0.73
91	1983: 07	0	1	1	1	0	0	1	3		-0.85
92	1983: 08	0	1	1	1	0	0	1	3		-0.78
93	1983: 09	0	1	1	1	0	0	1	3		-0.92
94	1983: 10	0	1	1	1	0	0	1	3		0.59
95	1983: 11	0	1	1	0	0	1	1	3		0.90
96	1983: 12	0	1	1	0	1	0	1	3		0.41

#	Date									+	value
97	1984: 01	0	0	0	0	1	0	0	6		0.58
98	1984: 02	0	1	0	0	1	0	0	5		0.20
99	1984: 03	0	0	0	0	1	0	0	6		-0.93
100	1984: 04	0	0	1	0	0	0	0	6		-1.16
101	1984: 05	0	0	1	1	0	0	0	5		-0.76
102	1984: 06	1	1	1	1	0	0	0	3		-0.57
103	1984: 07	1	1	1	1	0	0	0	3		-1.04
104	1984: 08	1	1	1	1	0	0	1	2		-0.55
105	1984: 09	1	1	1	1	1	0	1	1	+	-0.21
106	1984: 10	1	1	1	1	0	1	1	1	+	-0.20
107	1984: 11	1	1	1	1	0	1	1	1	+	0.33
108	1984: 12	0	1	0	1	0	1	1	3		0.25
109	1985: 01	0	0	0	0	0	1	0	6		0.01
110	1985: 02	0	0	0	0	0	1	0	6		0.73
111	1985: 03	0	0	0	0	0	1	0	6		0.45
112	1985: 04	0	0	0	0	0	1	0	6		-0.08
113	1985: 05	0	0	0	0	0	0	0	7		-0.15
114	1985: 06	0	0	0	0	0	0	0	7		-0.51
115	1985: 07	0	1	0	0	0	0	0	6		-0.61
116	1985: 08	0	1	0	1	1	0	1	3		-0.34
117	1985: 09	0	1	1	1	1	0	1	2		-0.08
118	1985: 10	0	1	1	1	1	0	1	2		-0.11
119	1985: 11	0	1	1	1	1	0	1	2		0.19
120	1985: 12	0	1	1	1	1	0	1	2		0.70
121	1986: 01	0	1	0	1	1	0	1	3		-0.04
122	1986: 02	0	0	1	1	1	0	0	4		-0.74
123	1986: 03	0	0	1	1	0	0	0	5		0.29
124	1986: 04	0	0	1	1	0	0	0	5		0.77
125	1986: 05	0	0	1	1	0	0	0	5		0.42
126	1986: 06	0	0	1	1	0	0	0	5		0.15
127	1986: 07	0	0	1	1	1	0	0	4		-0.08
128	1986: 08	0	0	0	0	1	0	1	5		0.76
129	1986: 09	0	0	0	0	1	1	1	4		1.90
130	1986: 10	0	0	0	0	1	1	1	4		1.56
131	1986: 11	0	0	0	0	1	1	1	4		0.98
132	1986: 12	0	0	0	0	1	1	1	4		0.59
133	1987: 01	0	0	0	0	1	1	1	4		0.66
134	1987: 02	1	0	0	0	1	1	1	3		0.11
135	1987: 03	1	0	0	0	1	1	1	3		-0.43
136	1987: 04	1	0	1	1	1	0	1	2		0.14
137	1987: 05	1	1	1	1	0	0	1	2		-0.20
138	1987: 06	1	1	1	1	0	0	1	2		-0.76
139	1987: 07	1	1	1	1	0	0	0	3		-1.52
140	1987: 08	1	1	1	1	0	0	0	3		-1.48
141	1987: 09	1	1	1	1	1	0	1	1	+	-0.90
142	1987: 10	1	1	1	1	1	1	1	0	+	-0.13
143	1987: 11	0	1	1	1	1	1	1	1	+	-0.01
144	1987: 12	0	1	1	1	1	1	1	1	+	0.29
145	1988: 01	0	1	1	1	1	1	0	2		-0.01
146	1988: 02	0	0	0	1	1	1	0	4		-0.35
147	1988: 03	0	0	0	1	1	1	0	4		-0.84
148	1988: 04	0	0	0	1	0	0	0	6		-0.96
149	1988: 05	0	1	1	1	0	1	0	3		-0.69
150	1988: 06	0	1	1	1	0	1	0	3		-1.21
151	1988: 07	0	1	1	1	0	1	1	2		-0.68
152	1988: 08	0	1	1	1	0	1	1	2		-0.66
153	1988: 09	0	1	1	0	1	1	1	2		-0.13
154	1988: 10	0	1	0	0	1	1	1	3		0.27
155	1988: 11	0	1	0	0	1	1	1	3		0.04
156	1988: 12	0	0	0	0	1	1	1	4		0.15
157	1989: 01	0	0	0	0	1	1	1	4		1.38
158	1989: 02	0	0	0	0	1	1	1	4		1.41
159	1989: 03	0	0	0	0	1	0	0	6		1.13
160	1989: 04	0	0	0	0	1	0	0	6		0.62
161	1989: 05	0	0	0	0	1	0	1	5		0.21
162	1989: 06	0	0	0	0	1	0	1	5		0.34
163	1989: 07	0	0	0	0	1	0	1	5		-0.15
164	1989: 08	0	0	0	0	1	0	1	5		0.39
165	1989: 09	0	0	0	0	1	1	1	4		1.19
166	1989: 10	0	0	0	0	1	1	1	4		1.47
167	1989: 11	0	0	0	0	1	1	1	4		1.20
168	1989: 12	0	0	0	0	1	1	1	4		1.18
169	1990: 01	0	0	0	0	1	1	1	4		0.97
170	1990: 02	0	0	0	0	0	0	0	7		1.60
171	1990: 03	0	0	0	0	0	0	0	7		1.15
172	1990: 04	0	0	0	0	0	0	0	7		0.92
173	1990: 05	0	0	0	0	0	0	0	7		0.21
174	1990: 06	0	0	0	0	0	0	0	7		-0.44
175	1990: 07	0	0	0	0	1	1	0	5		0.74
176	1990: 08	0	0	0	0	1	1	1	4		1.63
177	1990: 09	0	0	0	0	1	1	1	4		1.64
178	1990: 10	0	0	0	0	1	1	1	4		2.25
179	1990: 11	0	0	0	0	1	1	1	4		1.24
180	1990: 12	0	0	0	0	1	1	1	4		0.20
181	1991: 01	0	0	0	0	1	0	0	6		0.04
182	1991: 02	0	0	0	1	0	1	0	5		-1.16
183	1991: 03	0	0	0	1	0	0	0	6		-1.38
184	1991: 04	0	0	1	1	0	0	0	5		-1.40
185	1991: 05	0	0	1	1	0	0	0	5		-1.60
186	1991: 06	0	0	1	1	0	0	0	5		-1.14
187	1991: 07	0	0	1	1	1	0	0	4		-1.10
188	1991: 08	0	0	0	1	1	0	0	5		1.12
189	1991: 09	0	0	0	1	1	0	1	4		2.63
190	1991: 10	0	0	0	0	1	1	1	4		3.04
191	1991: 11	0	0	0	0	1	1	1	4		2.86
192	1991: 12	0	0	0	0	1	1	1	4		2.60
193	1992: 01	0	0	0	0	1	1	0	5		1.84
194	1992: 02	0	0	0	0	1	1	0	5		0.95
195	1992: 03	0	0	0	0	1	0	0	6		-0.58
196	1992: 04	0	0	0	1	0	0	0	6		-0.88
197	1992: 05	0	0	1	1	0	0	0	5		-1.70
198	1992: 06	0	1	1	1	0	0	0	4		-2.75
199	1992: 07	0	1	1	1	0	0	0	4		-1.21
200	1992: 08	0	1	1	1	0	0	0	4		0.10
201	1992: 09	0	1	1	1	0	0	0	4		1.72
202	1992: 10	0	1	1	1	1	0	1	2		2.35
203	1992: 11	0	1	1	1	1	0	1	2		2.47
204	1992: 12	0	0	1	1	0	0	0	5		1.76
205	1993: 01	0	0	0	1	0	0	0	6		2.04
206	1993: 02	1	0	0	1	0	0	0	5		1.04
207	1993: 03	1	0	0	1	0	0	0	5		0.21
208	1993: 04	1	0	0	1	0	0	0	5		-0.74
209	1993: 05	1	0	1	1	0	0	0	4		-0.46
210	1993: 06	1	1	1	1	0	0	0	3		-1.05

211	1993: 07	1	1	1	1	0	0	0	3	-1.60
212	1993: 08	0	1	1	1	0	0	0	4	-1.00
213	1993: 09	0	0	0	1	0	0	1	5	-0.20
214	1993: 10	0	0	0	1	0	1	1	4	0.14

| 215 | 1993: 11 | 0 | 0 | 0 | 1 | 0 | 1 | 1 | 4 | 0.52 |
| 216 | 1993: 12 | 0 | 0 | 1 | 1 | 0 | 1 | 1 | 3 | -0.12 |

6.4. Collective Behavior of Premonitory Trends: A Hypothetical Prediction Algorithm

Keilis-Borok et al. (2003) consider how the approach of a homicide surge is reflected in the *collective* behavior of the trends. The simplest description of this behavior is $\Delta(m)$ - the number of *non-premonitory* trends at a given month m. If the identification of premonitory trends is correct then $\Delta(m)$ should be low in the proximity of an *SHS*. By definition $\Delta(m)$ is the number of zeros in the binary code of the monthly situation. This is the so-called "Hamming distance" between that code and the code of the "pure" premonitory situation, $\{1,1,1,1,1,1,1\}$ when all seven trends listed in Table 26 are premonitory (Gvishiani and Kosobokov, 1981; Lichtman and Keilis-Borok, 1989; Keilis-Borok et al., 2000).

The values of $\Delta(m)$ are given in Table 27. Fig. 38 shows the change of $\Delta(m)$ with time. The value of $\Delta(m)$ may vary from 0 to 7 but the minimal observed value is 1; the corresponding lines in Table 27 are marked by "+". This value appears within 4 to 11 months before an *SHS* and at no other time. An examination of the temporal change of $\Delta(m)$ in Table 27 suggests the following hypothetical prediction algorithm:

An alarm is declared for 9 months each time when $\Delta(m) \leq D$ for two consecutive months (regardless of whether these two months belong or not to an already declared alarm).

Possible outcomes of such a prediction are illustrated in Fig. 34. The condition $\Delta(m) \leq D$ means, by definition, that D or less trends are not premonitory at the month m. A count of $\Delta(m)$ in Table 27 suggests to take $D = 1$. A waiting period of 9 months is introduced because the premonitory trends do not appear right before an *SHS*. The requirement that this condition holds two months in a row makes a prediction more reliable and reduces the total duration of alarms.

The alarms obtained by this algorithm are shown in Fig. 38 by the red bars. The total duration of these alarms is 30 months, representing 14 percent of all months considered. In real prediction that score would be quite satisfactory.

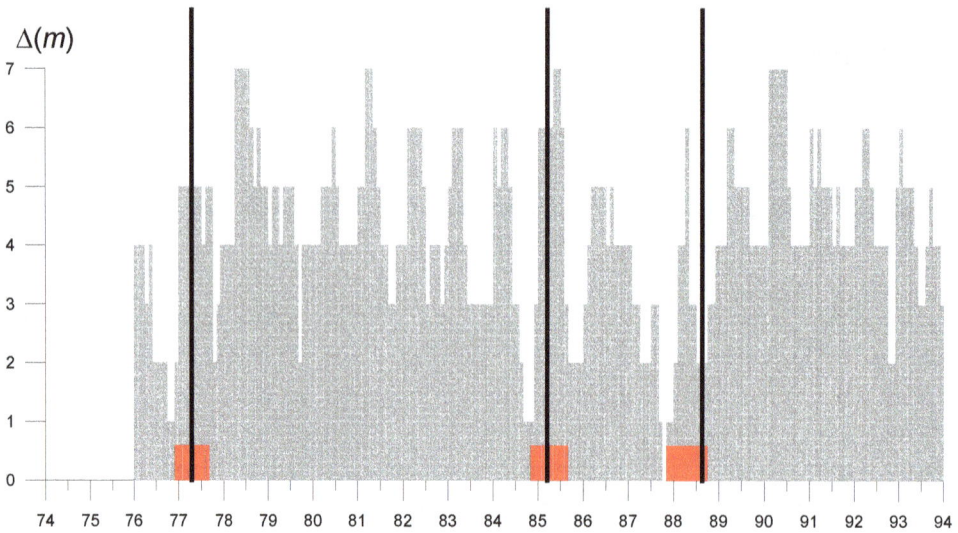

Figure 38. Homicide surges and alarms determined by the prediction algorithm. Start of a homicide surge is shown by the vertical line. Function $\Delta(m)$ is the number of crime statistics *not* showing premonitory trends at a month m. Alarms (shown by red bars) are declared for 9 months, when $\Delta(m) \leq 1$ during two consecutive month. Adjustable parameters correspond to version 10 of the algorithm (see Table 28).

6.5. Stability of Prediction (Sensitivity Analysis)

Inevitably in lieu of a set of fundamental equations for crime dynamics there is a certain freedom in the retrospective ad hoc choice of adjustable elements: the types of crimes considered, numerical parameters, such as percentiles Q^C, etc. An algorithm thus developed makes sense only if it is not too sensitive to variation of these choices; as Enrico Fermi put it, *"with four exponents I can fit an elephant"*.

To explore that sensitivity Keilis-Borok et al. (2003) repeat the prediction with different sets of the kinds of crime considered and with different values for the numerical parameters. These sets are described in Table 28. The outcomes of prediction are compared on the error diagrams (Fig. 39). Molchan (1997) has introduced such diagrams as a tool for evaluating prediction methods and optimizing disaster preparedness. Their application to research in prediction of recessions and unemployment are described in (Keilis-Borok et al, 2000, 2005).

The "basic" variant (Section 6.4) is # 10 in Table 28. The variations considered are discussed below.

Variation of the percentiles Q^C, defining discretization thresholds (# 8, 9, 11, 12). Lowering them, the total duration of alarms is obviously increased, but the results of prediction do not change much and remain acceptable.

Using only two kinds of crimes (# 12) comparable results are obtained. However it would be risky to make advance prediction with only two indicators.

Table 28. Variation of the adjustable elements

Crime type		Variants											
		1	2	3	4	5	6	7	8	9	10	11	12
Value of D*		3	5		8	4		3	1				0
		Trend and percentile											
Rob	PR**	upward			downward								
	Q^C, %	33.3	25.0	20.0	80.0	66.7						50.0	
FRob	PR				downward								
	Q^C, %				80.0	66.7						50.0	
KCIR	PR				downward								
	Q^C, %				80.0	66.7					50.0		66.7
ODWR	PR	upward			downward								
	Q^C, %	33.3	25.0	20.0	80.0	66.7	87.5						
SAR	PR	upward			downward								
	Q^C, %	33.3	25.0	20.0	80.0	66.7							
A	PR	upward											
	Q^C, %	33.3	25.0	20.0	20.0	33.3							
FA	PR	upward											
	Q^C, %	33.3	25.0	20.0	20.0	33.3			50.0				
KCIA	PR	upward											
	Q^C, %	33.3	25.0	20.0	20.0	33.3			50.0				
ODWA	PR	upward											
	Q^C, %	33.3	25.0	20.0	20.0	33.3							
AIA	PR	upward											
	Q^C, %	33.3	25.0	20.0	20.0	33.3							
UNFE	PR				upward				upward				
	Q^C, %				20.0	33.3			33.3	50.0			33.3
AFE	PR				upward								
	Q^C, %				20.0	33.3							

*The values that give relatively best performance for that variant.
**PR is an abbreviation for the premonitory trend.

The limits of acceptable variations are reached in the other variants (# 1 – 7). Keilis-Borok et al. (2003) tried to find a premonitory *rise* of robberies, simultaneous with rise of other crimes and consider other kinds of crime; in all variants its performance remains unacceptable.

For advance prediction variants 8 – 11 might be used in parallel. Such parallel predictions might better suit the needs of a decision-maker, determining possible disaster preparedness measures (Kossobokov et al., 2000b; Zaliapin et al., 2003b).

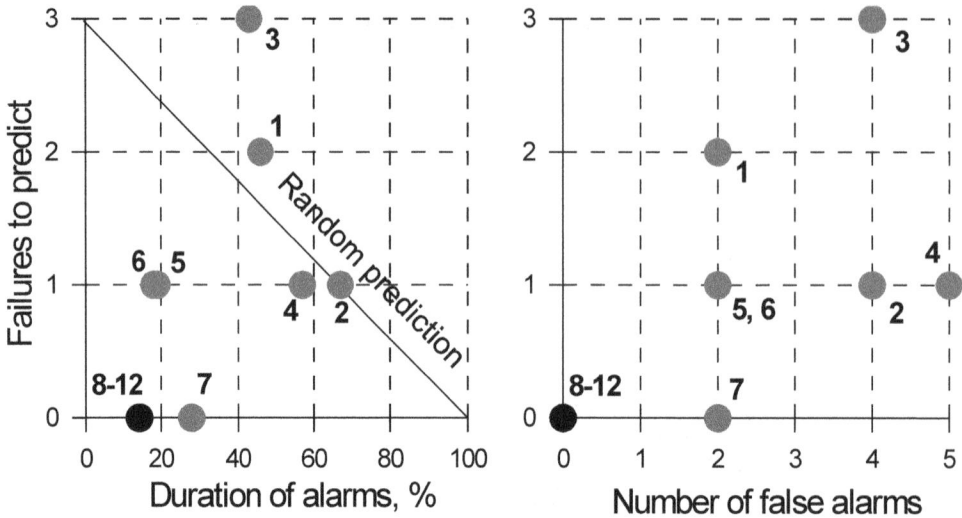

Figure 39. Error diagram. Numbers near the dots identify the variant of the algorithm in Table 28. Black dots show the variants suggested for advance prediction. See further explanations in Section 6.5.

6.6. Applications to Independent Data

Here the algorithm is tested by application to "out of sample" data not used in its development. Such tests are always necessary to validate and/or improve a prediction algorithm. Such a test is possible since the algorithm is self-adaptive: the thresholds $T^c(Q^c)$ are not fixed but are adapted to crime statistics, as the percentile of a level Q^c.

Los Angeles, 1994-2002. So far the data source (Carlson, 1998) covering the years 1975 – 1993 was used. To extend the analysis past 1993, there is the data of the LAPD Information Technology

Division, covering the time period from January 1990 to May 2002. Comparing the data for the overlapping three years it is found that they are reasonably close, particularly after smoothing.

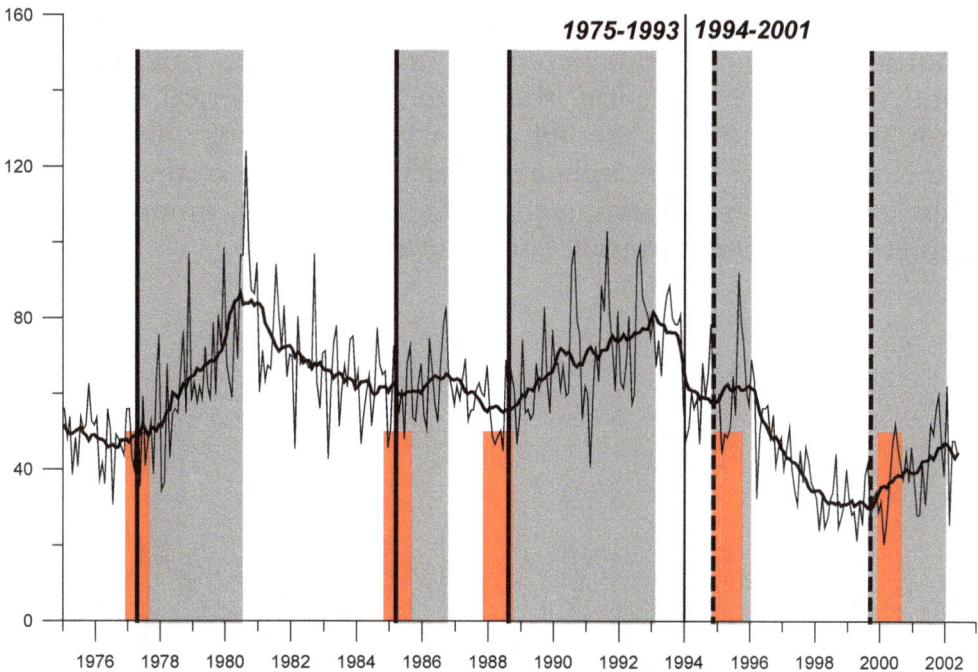

Figure 40. Performance of prediction algorithm through 1975-2002. Data from the National Archive of Criminal Justice Data (Carlson, 1998) for 1975 – 1993 have been used to develop the algorithm. It was then applied to the data from the Data Bank of the Los Angeles Police Department (LAPD Information Technology Division) for subsequent 9 years. Notations are the same as in Fig. 36. Dashed vertical lines indicate *SHS* episodes that occurred after 1993.

Fig. 40 shows the homicide rates through the whole period from 1975 to May 2002. Two *SHS* episodes are identified in the later period 1994-2001. They are indicated in Fig. 40 by dashed vertical lines. The first episode is captured by an alarm, which starts in the month of *SHS* without a lead time. The second episode is missed, since an alarm has started two months after it. This error needs to be put on the record; nevertheless the prediction remains informative: during these two months homicide rose by only a few percent, giving no indication that a lasting homicide surge had started.

New York City. Fig. 41 shows the monthly total homicide rates in New York City, per 7 million inhabitants of the city. Two *SHS* episodes (02:1978 and 02:1985) were identified. The prediction algorithm gives two alarms, as shown in Fig. 41 by red bars. One of them predicts the second *SHS*, while the first one is missed. Another alarm is considered as a false one; this has to be confirmed by processing the data for the period after 1993. Though the failure to predict and a false alarm are disappointing, the results as a whole appear to be useful: one of the two *SHS* is captured by alarms lasting together 21 months, amounting to 10 percent of the time interval considered.

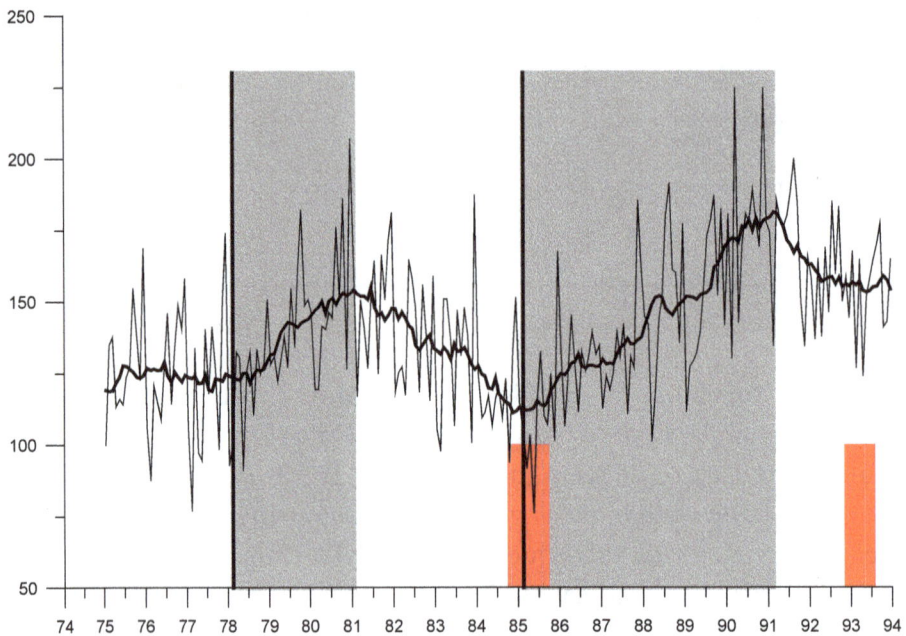

Figure 41. Application of the prediction algorithm to New York City. Notations are the same as in Fig. 36. Data are taken from the National Archive of Criminal Justice Data (Carlson, 1998). Homicide statistics is shown per 7,000,000 of inhabitants.

6.7. On a More Precise Prediction

A conjecture that could be worth exploring in the future is outlined here. Two consecutive patterns of the crimes considered have been observed. The first one precedes an *SHS* with a lead time of 4 to 11 months; it is formally defined and explored in Sections 4, 5. The second pattern emerges with a shorter lead time, promising a more accurate prediction of the time of an incipient *SHS*.

A distinctive trait of the second pattern is a steep simultaneous rise of the different types of robberies. Let us replace this pattern by a less specific one that is more broadly defined: the absence of a steep decline. By definition, that pattern will be captured by the zeros in the first four columns of Table 27. Counting them, one can find that three or more emerge within 6 months before each *SHS*. This result suggests the following second approximation to the prediction algorithm described above. Consider the period of alarm declared by the algorithm; let us call it "the first phase alarm". Within that period *a "second phase alarm" is declared for 6 months after the first month when* $\Delta_1(m) \leq 1$. Here $\Delta_1(m)$ is the number of ones in the codes of the robberies (the first four columns in Table 27). In the absence of the first-phase alarm the second one is not declared.

Alarms obtained by this rule are shown in Fig. 42. The alarms become much shorter; their total duration drops to 18 months, that is, from 14 percent to 8 percent of all the months considered. Even better results could potentially be obtained by directly capturing *a rise* of robberies, but this requires weekly, if not daily, crime rates (since the lead time of the rise in robberies is relatively short).

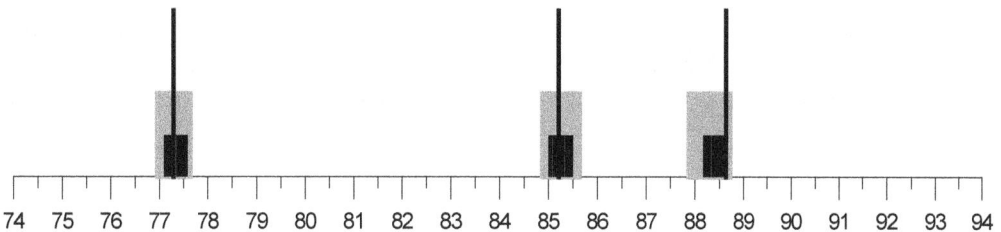

74 75 76 77 78 79 80 81 82 83 84 85 86 87 88 89 90 91 92 93 94

Figure 42. Possible reduction of duration of alarms. Vertical lines – starting points of a homicide surge (*SHS*). Grey bars – alarms obtained by the suggested algorithm. Black bars – alarms obtained in a hypothetical second approximation.

Using the trend of homicides themselves might provide a similar possibility. Values of the function $K^H(m\text{-}12,\ m)$, which estimates that trend (see Section 6.3) are given in Table 27 (column K^H). Within each alarm one can see the months when $K^H(m\text{-}12,\ m) > 0$. Starting alarms at these months, it would be possible to reduce further the duration of alarms without having an additional failure to predict.

6.8. Conclusions

The conclusions of Keilis-Borok et al. (2003) might be summed up as follows. Crime statistics in the city of Los Angeles for the period 1975 – 2001 was analyzed, exploring the possibility of anticipating a turn of the homicide rate from decline to a surge. It has been found that within 4 to 11 months such a turn is preceded by a specific pattern of the crime statistics: both burglaries and assaults escalate, while robberies decline, along with the homicides themselves. Both changes, escalation and decline, are not monotonic, but occur sporadically, each lasting 2-6 months.

Based on this pattern a prediction algorithm is formulated, giving it a robust and unambiguous definition. It is self-adapting to average crimes statistics, so that one could apply it to New York City as well as Los Angeles. The major limitation of this study is that, as is inevitable for an initial study, only a small number of homicide surges was available for analysis. The algorithm remains hypothetical until it is validated by advance prediction. It is encouraging, however, that those predictions are stable as to variations in the adjustable elements of the algorithm.

Closer to the surge of homicides, the robberies also turn from decline to rise. This indicates the possibility of a second approximation to the prediction, with twice the accuracy (that is with a twofold reduction in the duration of alarms).

The analysis given above captures the consecutive escalation of different crimes: first – of burglaries and assaults only, then of robberies, then of homicides. That sequence, albeit hypothetical so far, seems natural, being in good accord with previous experience in the following areas.

(*i*) The sequence reflects a more general phenomenon, commonly known in law enforcement practice: a consecutive escalation of more and more severe crimes, signaling that a surge of major crimes is approaching. A quantitative definition of a specific manifestation of this phenomenon is given by Keilis-Borok et al. (2003). Similar escalation has been found in French suburban areas (Bui Trong, 2003).

(*ii*) The sequence is also in accord with a well-known "universal" feature of many hierarchical complex systems: the rise of permanent background activity ("static") of the system culminated by a fast major change – a "critical transition". This feature happens to be common for different physical and socio-economic systems. It is reproduced by the "universal" models of hierarchical complex systems, such as those developed in theoretical physics (e.g. Yamashita and Knopoff, 1992; Gell-Mann, 1994; Holland, 1995; Newman et al., 1995; Turcotte, 1997; Allègre et al., 1998; Shnirman, and Blanter, 1998; Gabrielov et al., 2000b; Rundle et al., 2000; Sornette, 2000; Turcotte et al., 2000; Zaliapin et al., 2003b).

This feature was also observed in many very different real world systems. For example, in earthquake-prone regions the "static" includes background seismicity. Premonitory escalation of seismic activity is a well-known precursor to major earthquakes, which is used in many earthquake prediction algorithms (e.g. Keilis-Borok and Kossobokov, 1990; Kossobokov and Carlson, 1995; Newman et al., 1995; Keilis-Borok, 2002). In economy, the "static" includes various macroeconomic indicators. Their premonitory escalation has been successfully used in the prediction of recessions and surges of unemployment (Keilis-Borok et al., 2000, 2005).

The results of Keilis-Borok et al. (2003) are also in accord with a distinctive common trait of precursors established in many of these studies: premonitory evolution of background activity is not monotonic, but realised sporadically, in a sequence of relatively short intermittent changes.

The universality of premonitory phenomena is limited and cannot be taken for granted in studying any specific system. Nevertheless, it is worth exploring in crime dynamics other known types of premonitory patterns, e.g. the clustering of background activity and the rise of the correlation range (Keilis-Borok, 2002; Gabrielov et al., 2000b; Shebalin et al., 2000).

What is the place of the study of Keilis-Borok et al. (2003) in the broad field of prediction of crime dynamics? Specific features of this approach might be summed up as follows.

Keilis-Borok et al. (2003) are trying to predict not the whole dynamics of homicides, but only the relatively rare phenomena - episodes of *SHS*.

Accordingly, Keilis-Borok et al. (2003) are looking for a quantitative and precisely defined prediction algorithm of the *"yes or no"* variety: at any moment of time such an algorithm would indicate whether or not such an episode should be expected within a fixed time interval.

The analysis of Keilis-Borok et al. (2003) is intentionally robust, which makes the prediction algorithm more reliable and applicable in different circumstances. In this case the performance of the algorithm did not change through the period considered even though Los Angeles has witnessed many changes relevant to crime over this period. This stability is achieved at a price, however, in that the time of a homicide surge is predicted with limited accuracy and the duration of a surge even more so.

The approach of Keilis-Borok et al. (2003) – a heuristic "technical" analysis - is not competing with but complementary to the cause-and-effect "fundamental" analysis. The cause that triggered a specific homicide surge is usually known, at least in retrospect. This might be, for example, a rise in drug use, a rise in unemployment, a natural disaster etc. However, that does not render predictions considered in this study redundant. On the contrary, the approach might predict an unstable situation when a homicide surge might be triggered, thus enhancing the reliability of cause-and-effect predictions.

It is encouraging for further studies in this direction that only a small part of the relevant and available data that can be incorporated in the analysis were used here. Among these are other types of crimes (Bursik et al., 1990), economic and demographic indicators (Messner, 1983) and the territorial distribution of crimes. It seems worthwhile to try the same approach with other targets of prediction – e.g. surges of all violent crimes; and to other areas, e.g. separate Bureaus of the city of Los Angeles, or to other major cities. In a broader scheme of things, the analysis of Keilis-Borok et al. (2003) discriminates stable situations from unstable, where the risk of different disasters is higher.

AFTERWORD:

Lessons Learned and Outlook for the Future

While the prediction algorithms described here still need to be improved, the following findings seem promising.

1. In each case, the prediction rules are uniform, transcending the immense complexity of the processes considered, the diversity of prediction targets, and the changes of circumstances in time. For the earthquakes, premonitory seismicity patterns happen to be similar for microcracks in laboratory samples to the largest earthquakes of the world, spanning the energy range from 10^{-1} erg to 10^{26} erg. Similar patters also likely apply to the ruptures in a neutron star, 10^{41}erg. For recessions and unemployment, the similarities between them overcome the changes of economy since 1962; for FAUs, their similarities also overcome the differences between the countries.

2. Our predictions are always based on the routinely available data — global catalogs of earthquakes, major macroeconomic indicators, etc.

3. In each case predictability is achieved by extremely robust analysis. It is interesting to quote how an economist with excellent track record in recessions' prediction describes his impressions on such a robustness:

"Prediction of recessions... requires fitting non-linear, high-dimensional models to a handful of observations generated by a possibly non-stationary economic environment....The evidence presented here suggests that these simple binary transformations of economic indicators have significant predictive content comparable to or, in many cases, better than that of more conventional models."

J. Stock (from Keilis-Borok et. al., 2000)

The accuracy of prediction is still limited. However, only a small part of relevant models and available data has been used so far. A wealth of possibilities for better prediction remains unexplored.

REFERENCES

Aki K (1996) Scale dependence in earthquake phenomena and its relevance to earthquake prediction. Proc Natl Acad Sci USA 93: 3740-3747.

Allègre CJ, Le Mouël J-L, Ha Duyen C, Narteau C (1995) Scaling organization of fracture tectonics (SOFT) and earthquake mechanism. Phys Earth Planet Inter, 92: 215-233.

Allègre CJ, Shebalin P, Le Mouel J-L, Narteau C (1998) Energetic balance in scaling organization of fracture tectonics. Phys Earth Planet Inter 106: 139-153.

Barenblatt GM, Keilis-Borok VI, Monin AS (1983) Filtration model of earthquake sequence. Trans (Doklady) Acad Sci SSSR 269: 831-834.

Barenblatt G (1996) Scaling, Self-Similarity, and Intermediate Asymptotics, Cambridge University Press, Cambridge.

Blanter EM, Shnirman MG, Le Mouël J-L, Allègre CJ (1997) Scaling laws in blocks dynamics and dynamic self-organized criticality. Phys Earth Planet Inter, 99: 295-307.

Bongard MM (1970) Pattern Recognition. Rochelle Park, N.J.: Hayden Book Co., Spartan Books.

Bowman DD, Ouillon G, Sammis G.G, Sornette A, Sornette D (1998) An observational test of the critical earthquake concept. J Geophys Res 103: 24359-24372.

Bursik RJ, Jr, Grasmick HG, Chamlin MB (1990) The effect of longitudinal arrest patterns on the development of robbery trends at the neighborhood level. Criminology 28(3): 431-450.

Bufe CG, Varnes DJ (1993) Predictive modeling of the seismic cycle of the greater San Francisco Bay region. J Geophys Res 98: 9871–9883

Bui Trong L (2003) Risk of collective youth violence in French suburbs. A clinical scale of evaluation, an alert system. In: Beer T, Ismail-Zadeh A (eds) Risk Science and Sustainability. Kluwer Academic Publishers, Dordrecht-Boston-London (NATO Science Series. II. Mathematics, Physics and Chemistry – Vol. 112), pp 199-221.

Burridge R, Knopoff L (1967) Model and theoretical seismicity. Bull Seism Soc Am 57: 341-371.

Caputo M, Gasperini P, Keilis-Borok V, Marcelli L, Rotwain I (1977) Earthquake's swarms as forerunners of strong earthquakes in Italy. Annali di Geofisica XXX(3/4): 269-283.

Caputo M, Console R, Gabrielov AM, Keilis-Borok VI, Sidorenko TV (1983) Long-term premonitory seismicity patterns in Italy. Geophys J R Astron Soc 75: 71-75.

Carlson SM (1998) Uniform Crime Reports: Monthly Weapon-specific Crime and Arrest Time Series, 1975-1993 (National, State, and 12-City Data). ICPSR 6792, Inter-university Consortium for Political and Social Research, P.O. Box 1248, Ann Arbor, Michigan 48106.

Crutchfield JP, Farmer JD, Packard NH, Shaw RS (1986) Chaos. Sci Am 255:46–57

Davis C, Goss K, Keilis-Borok V, Molchan G, Lahr P, Plumb C (2007) Earthquake Prediction and Tsunami Preparedness. Workshop on the Physics of Tsunami, Hazard Assessment Methods and Disaster Risk Management, 14-18 May 2007, Trieste: ICTP.

Davis CA, Keilis-Borok V, Molchan G, Shebalin P, Lahr P, Plumb C (2010) Earthquake prediction and disaster preparedness: Interactive analysis. Natural Hazards Review, ASCE 11(4): 173-184.

Davis C, Keilis-Borok V, Kossobokov V, Soloviev A (2012) Advance prediction of the March 11, 2011 Great East Japan Earthquake: A missed opportunity for disaster preparedness. International Journal of Disaster Risk Reduction, 1: 17-32, doi:10.1016/j.ijdrr.2012.03.001.

Engle RF, McFadden DL (1994) (eds) Handbook of Econometrics, volume 4. North-Holland Publishing Co., Amsterdam.

Farmer JD, Sidorowich J (1987) Predicting chaotic time series. Phys Rev Lett 59: 845.

Gabrielov A, Dmitrieva OE, Keilis-Borok VI, Kossobokov VG, Kuznetsov IV, Levshina TA, Mirzoev KM, Molchan GM, Negmatullaev SKh, Pisarenko VF, Prozoroff AG, Rinehart W, Rotwain IM, Shebalin PN, Shnirman MG, Shreider SYu (1986) Algorithm of Long-Term Earthquakes' Prediction. Centro Regional de Sismologia para America del Sur, Lima, Peru.

Gabrielov AM, Keilis-Borok VI, Jackson DD (1996) Geometric incompatibility in a fault system. Proc Natl Acad Sci USA 93: 3838-3842.

Gabrielov A, Keilis-Borok V, Zaliapin I, Newman WI (2000a) Critical transitions in colliding cascads. Phys Rev E 62, 237-249.

Gabrielov AM, Zaliapin IV, Newman WI, Keilis-Borok VI (2000b) Colliding cascade model for earthquake prediction. Geophys J Int 143(2): 427-437.

Gelfand IM, Guberman ShA, Keilis-Borok VI, Knopoff L, Press F, Ranzman IYa, Rotwain IM, Sadovsky AM (1976) Pattern recognition applied to earthquake epicenters in California. Phys Earth Planet Inter 11: 227-283.

Gell-Mann M (1994) The Quark and the Jaguar: Adventures in the Simple and the Complex. Freeman and Company, New York.

Ghil M (1994) Cryothermodynamics: The chaotic dynamics of paleoclimate. Physica D 77: 130-159.

Gorshkov A, Keilis-Borok V, Rotwain I, Soloviev A, Vorobieva I (1997) On dynamics of seismicity simulated by the models of blocks-and-faults systems. Annali di Geofisica XL(5): 1217-1232.

Gvishiani AD, Kossobokov VG (1981) Proc Ac Sci USSR. Phys Earth 2: 21-36.

Holland JH (1995) Hidden Order: How Adaptation Builds Complexity. Addison-Wesley, Reading (Mass).

Huang Y, Saleur H, Sammis C, Sornette D (1998) Precursors, aftershocks, criticality and self-organized criticality. Europhys Lett 41: 43-48.

IMF (1997) International Monetary Fund, International Financial Statistics, CD-ROM.

Jaumé SC, Sykes LR (1999) Evolving to wards a critical point: A review of accelerating seismic moment/energy release prior to large and great earthquakes. Pure Appl Geophys 155: 279-306.

Kantorovich LV, Keilis-Borok VI, Molchan GM (1974) Seismic risk and principles of seismic zoning. In: Seismic design decision analysis. Department of Civil Engineering, MIT, Internal Study Report 43.

Kadanoff LP (1976) Scaling, universality and operator algebras. In: Domb C, Green MS (eds) Phase Transitions and Critical Phenomena, Vol 5a

Keilis-Borok VI, Malinovskaya LN (1964) One regularity in the occurrence of strong earthquakes. J Geophys Res 69:3019–3024.

Keilis-Borok VI, Knopoff L, Rotwain IM (1980) Bursts of aftershocks, long-term precursors of strong earthquakes. Nature 283: 258-263.

Keilis-Borok VI, Press F (1980) On seismological applications of pattern recognition. In: Allegre CJ (ed) Source Mechanism and Earthquake Prediction Applications. Editions du Centre national de la recherché scientifique, Paris, pp 51-60.

Keilis-Borok V, Lamoreau R, Johnson C, Minster B (1982) Swarms of main shocks in Southern California. In: Rikitake T (ed) Earthquake Prediction Research. Elsevier, Amsterdam.

Keilis-Borok VI, Kossobokov VG (1984) A complex of long-term precursors for the strongest earthquakes of the world. In: Proceedings 27th Geological Congress, 61, Nauka, Moscow, pp 56–66.

Keilis-Borok VI, Kossobokov VG (1987) Periods of high probability of occurrence of the world's strongest earthquakes. In: Computational Seismology, 19, Allerton Press, pp 45–53.

Keilis–Borok VI (1990) (ed) Intermediate-term Earthquake Prediction: Models, Phenomenology, Worldwide Tests. Phys Earth Planet Inter 61, special issue 1-2: 1-144.

Keilis-Borok VI, Kossobokov VG (1990) Premonitory activation of earthquake flow: algorithm M8. Phys Earth Planet Inter 61: 73-83.

Keilis-Borok VI, Rotwain IM (1990) Diagnosis of time of increased probability of strongearthquak es in different regions of the world: algorithm CN. Phys Earth Planet Inter 61: 57-72.

Keilis-Borok VI, Lichtman AJ (1993) The self-organization of American society in presidential and senatorial elections. In: Kravtsov YuA (ed) Limits of Predictability. Springer-Verlag, Berlin-Heidelberg, pp 223-237.

Keilis-Borok VI (1994) Symptoms of instability in a system of earthquake-prone faults. Physica D 77: 193-199.

Keilis-Borok VI, Rotwain IM, Shebalin PN (1994) The special redistribution of low-level seismicity in focal zones of large earthquakes. In: Keilis-Borok VI (ed) Seismicity and Related Processes in the Environment, volume 1. Research and Coordinating Centre for Seismology and Engineering, Moscow, pp 44-48.

Keilis-Borok, V (1996) Intermediate-term earthquake prediction. Proc Natl Acad Sci USA 93: 3748-3755.

Keilis-Borok VI, Shebalin PN (1999) (eds) Dynamics of Lithosphere and Earthquake Prediction. Phys Earth Planet Inter 111, special issue 3-4: 179-330.

Keilis-Borok VI, Sorondo MS (2000) (eds) Science for Survival and Sustainable Development. The Proceedings of the Study-Week of the Pontifical Academy of Sciences, 12-16 March 1999. Pontificiae Academiae Scientiarvm Scripta Varia, 98, Vatican City.

Keilis-Borok V, Stock JH, Soloviev A, Mikhalev P (2000) Pre-recession pattern of six economic indicators in the USA. J Forecast 19:65–80.

Keilis-Borok VI (2002) Earthquake prediction: State-of-the-art and emerging possibilities. Annu Rev Earth Planet Sci 30: 1-33.

Keilis-Borok V, Shebalin P, Zaliapin I (2002) Premonitory patterns of seismicity months before a large earthquake: Five case histories in Southern California. Proc Natl Acad Sci USA 99: 16562-16567.

Keilis-Borok VI (2003) Basic science for prediction and reduction of geological disasters. In Beer, T. and Ismail-Zadeh, A. (eds.), Risk Science and Sustainability, Kluwer Academic Publishers, Dordrecht-Boston-London (NATO Science Series. II. Mathematics, Physics and Chemistry – Vol. 112), pp. 29-38.

Keilis-Borok VI, Gascon DJ, Soloviev AA, Intriligator MD, Pichardo R, Winberg FE (2003) On predictability of homicide surges in megacities. In: Beer T, Ismail-Zadeh A (eds) Risk Science and Sustainability. Kluwer Academic Publishers, Dordrecht-Boston-London (NATO Science Series. II. Mathematics, Physics and Chemistry – Vol. 112), pp 91-110.

Keilis-Borok VI, Soloviev AA (2003) (eds) Nonlinear Dynamics of the Lithosphere and Earthquake Prediction, Springer-Verlag, Berlin-Heidelberg

Keilis-Borok V, Shebalin P, Gabrielov A, Turcotte D (2004) Reverse tracing of short-term earthquake precursors. Phys Earth and Planet Inter 145(1-4):75-85.

Keilis-Borok VI, Soloviev AA, Allègre CB, Sobolevskii AN, Intriligator MD (2005) Patterns of macroeconomic indicators preceding the unemployment rise in Western Europe and the USA. Pattern Recognition 38(3):423-435.

Keilis-Borok V, Soloviev A, Intriligator M, Winberg F (2006) Current prediction of the increase in the unemployment rate in the U.S. E2-C2 Meeting, Perugia, 02-05 September 2006, Golf Hotel Colle della Trinita.

Keilis-Borok VI, Soloviev AA, Intriligator MD, Winberg FE (2008) Pattern of macroeconomic indicators preceding the end of an American economic recession. J. Pattern Recognition Res., 3(1):40-53.

Klein PhA, Niemira MP (1994) Forecasting financial and economic cycles. New York, Wiley.

Knopoff L, Levshina T, Keilis-Borok VI, Mattori C (1996) Increased long-range intermediate-term earthquake activity prior to strong earthquakes in California. J Geophys Res 101(B3): 5779-5796.

Kosobokov VG (1983). Recognition of the sites of strong earthquakes in East Central Asia and Anatolia by Hamming's method. In: Keilis-Borok VI and Levshin AL (eds) Mathematical models of the structure of the Earth and the earthquake prediction, Comput. Seismol., 14. Allerton Press, New York, pp. 78-82.

Kossobokov VG, Keilis-Borok VI, Smith SW (1990) Localization of intermediate-term earthquake prediction. J Geophys Res 95(B12): 19763-19772.

Kossobokov VG, Carlson JM (1995) Active zone size vs. activity: A study of different seismicity patterns in the context of the prediction algorithm M8. J Geophys Res 100: 6431-6441.

Kossobokov VG, Keilis-Borok VI, Cheng B (2000a) Similarities of multiple fracturingon a neutron star and on the Earth. Phys Rev E 61(4): 3529-3533.

Kossobokov VG, Keilis-Borok VI, Turcotte DL, Malamud BD (2000b) Implications of a statistical physics approach for earthquake hazard assessment and forecasting. Pure Appl Geophys 157(11-12): 2323-2349.

Kossobokov V, Shebalin P (2003) Earthquake prediction. In: Keilis-Borok VI, Soloviev AA (eds) Nonlinear Dynamics of the Lithosphere and Earthquake Prediction. Springer-Verlag, Berlin-Heidelberg, pp 141-207.

Kravtsov YuA (1993) (ed) Limits of Predictability. Springer-Verlag, Berlin-Heidelberg.

Kuznetsov IV, Keilis-Borok VI (1997) The interrelation of earthquakes of the Pacific seismic belt. Trans (Doklady) Russ Acad Sci, Earth Sci Sect 355A(6): 869–873.

Levshina T, Vorobieva I (1992) Application of algorithm for prediction of a strong repeated earthquake to Joshua Tree and Landers. In: Fall Meeting AGU, Abstracts, p 382.

Lichtman AJ, Keilis-Borok VI (1989) Aggregate-level analysis and prediction of midterm senatorial elections in the United States, 1974-1986. Proc Natl Acad Sci USA 86(24): 10176-10180.

Lichtman AJ (1996) The Keys to the White House. Madison Books, Lanham

Lichtman AJ (2000) The Keys to the White House. Lexington Books Edition, Lanham

Lichtman AJ (2005) The Keys to the White House: Forecast for 2008. Foresight: The International Journal of Applied Forecasting. 3: 5-9.

Ma Z, Fu Z, Zhang Y, Wang C, Zhang G, Liu D (1990) Earthquake Prediction: Nine Major Earthquakes in China. Springer-Verlag, New York

Mason IB (2003) Binary events. In: Jolliffe IT, Stephenson DB (eds.) Forecast Verification. A Practioner's Guide in Atmospheric Science, Wiley and Sons Ltd, Chichester, pp 37-76.

Messner SF (1983) Regional differences in the economic correlates of the urban homicide rate. Criminology 21(4): 477-488.

Mogi K (1968) Migration of seismic activity. Bull Earth Res Inst Univ Tokyo 46(1): 53-74.

Molchan GM (1990) Strategies in strong earthquake prediction. Phys Earth Planet Inter 61:84-98.

Molchan GM, Dmitrieva OE, Rotwain IM, Dewey J (1990) Statistical analysis of the results of earthquake prediction, based on bursts of aftershocks. Phys. Earth Planet. Inter., 61: 128-139.

Molchan GM (1991) Structure of optimal strategies of earthquake prediction. Tectonophysics 193:267-276.

Molchan GM (1994) Models for optimization of earthquake prediction. In: Chowdhury DK (ed) Computational Seismology and Geodynamics, Vol 1. Am Geophys Un, Washington, DC, pp 1-10.

Molchan GM (1997) Earthquake prediction as a decision-making problem. Pure Appl Geophys 149: 233–237

Molchan GM (2003) Earthquake Prediction Strategies: A Theoretical Analysis. In: Keilis-Borok VI, Soloviev AA (eds) Nonlinear Dynamics of the Lithosphere and Earthquake Prediction, Springer-Verlag, Berlin-Heidelberg, pp 209–237.

Molchan G, Keilis-Borok V (2008) Earthquake prediction: Probabilistic aspect. Geophys J Int 173(3): 1012-1017.

Mostaghimi M, Rezayat F (1996) Probability forecast of a downturn in U.S. economy using classical statistical theory. Empirical Economics 21: 255-279.

Newman WI, Gabrielov AM, Turcotte DL (1994) (eds) Non-linear Dynamics and Predictability of Geophysical Phenomena, Geophysical Monograph Series, IUGG-Am Geophys. Union, Washington, DC.

Newman WI, Turcotte DL, Gabrielov AM (1995) Log-periodic behavior of a hierarchical failure model with application to precursory seismic activation. Phys Rev E. 52: 4827-4835.

OECD (1997) Main Economic Indicators: Historical Statistics 1960-1996. Paris, CD-ROM.

Panza GF, Soloviev AA, Vorobieva IA (1997) Numerical modeling of block-structure dynamics: Application to the Vrancea region. Pure Appl Geophys 149: 313-336.

Pepke GF, Carlson JR, Shaw BE (1994) Prediction of large events on a dynamical model of fault. J Geophys Res 99: 6769–6788.

Pollitz FF, Burgmann R, Romanowicz B (1998). Science 280: 1245-1249.

Press F, Briggs P (1975) Earthquakes, Chandler wobble, rotation and geomagnetic changes: a pattern recognition approach. Nature, 256: 270-273.

Press F, Briggs P (1977) Pattern recognition applied to uranium prospecting. Nature, 268: 125-127.

Press F, Allen C (1995) Patterns of seismic release in the southern California region. J Geophys Res 100(B4): 6421–6430.

Prozorov AG (1975) Changes of seismic activity connected to large earthquakes. In: Keilis-Borok VI (ed) Interpretation of Data in Seismology and Neotectonics, Comput Seismol 8. Nauka, Moscow, pp 71-82.

Richter C (1964) Comment on the paper "One Regularity in the Occurrence of StrongEarthquak es" by Keilis-Borok, V.I. and Malinovskaya, L.N. J Geophys Res 69: 3025.

Romanowicz B (1993) Spatiotemporal patterns in the energy-release of great earthquakes. Science 260: 1923–1926.

Rotwain I, Keilis-Borok V, Botvina L. (1997) Premonitory transformation of steel fracturing and seismicity. Phys Earth Planet Int 101: 61-71.

Rundkvist DV, Rotwain IM (1996) Present-day geodynamics and seismicity of Asia minor. In: Chowdhury DK (ed) Computational Seismology and Geodynamics, Vol 3. Am Geophys Un, Washington, DC, pp 130-149.

Rundle BJ, Turcotte DL, Klein W, (2000) (eds) Geocomplexity and the Physics of Earthquakes. Washington, DC: Am Geophys Union.

Sammis CG, Sornett D, Saleur H (1996) Complexity and earthquake forecasting. In: Rundle JB, Turcotte DL, Klein W (eds) SFI Studies in the Science of Complexity, vol. XXV. Addison-Wesley, Reading, Mass.

Shebalin P, Girardin N, Rotwain I, Keilis-Borok V, Dubois J (1996) Local overturn of active and non-active seismic zones as a precursor of large earthquakes in Lesser Antillean Arc. Phys Earth Planet Inter 97: 163-175.

Shebalin PN, Keilis-Borok VI (1999) Phenomenon of local "seismic reversal" before strong earthquakes. Phys Earth Planet Inter 111(3-4): 215-227.

Shebalin P, Zaliapin I, Keilis-Borok V (2000) Premonitory raise of the earthquakes' correlation range: Lesser Antilles. Phys Earth Planet Inter 122(3-4): 241-249.

Shebalin P, Keilis-Borok V, Zaliapin I, Uyeda S, Nagao T, Tsybin N (2004) Advance short-term prediction of the large Tokachi-oki earthquake, September 25, 2003, M = 8.1. A case history. Earth, Planets and Space 56(8): 715-724.

Shebalin P, Keilis-Borok V, Gabrielov A, Zaliapin I, Turcotte D (2006) Short-term earthquake prediction by reverse analysis of lithosphere dynamics. Tectonophysics 413: 63-75.

Shnirman MG, Blanter EM (1998) Self-organized criticality in a mixed hierarchical system. Phys Rev Letters 81: 5445-5448.

Shnirman M, Blanter E (2003) Hierarchical models of seismicity. In: Keilis-Borok VI, Soloviev AA (eds) Nonlinear Dynamics of the Lithosphere and Earthquake Prediction. Springer-Verlag, Berlin-Heidelberg, pp 37-69.

Soloviev A, Ismail-Zadeh A (2003) Models of dynamics of block-and-fault systems. In: Keilis-Borok VI, Soloviev AA (eds) Nonlinear Dynamics of the Lithosphere and Earthquake Prediction. Springer-Verlag, Berlin-Heidelberg, pp 71-139.

Sornette D, Sammis CG (1995) Complex critical exponents from renormalization group theory of earthquakes: Implications for earthquake predictions. J Phys I France 5: 607-619.

Sornette D (2000) Critical Phenomena in Natural Sciences: Chaos, Fractals, Self-organization, and Disorder. Concept and Tools. Springer, Berlin.

Stock JH, Watson MW (1989) New indexes of leading and coincident economic indicators. In: NBER Macroeconomics Annual, pp 351-394.

Stock JH, Watson MW (1993) A procedure for predicting recessions with leading indicators. In: Stock JH, Watson MW (eds) Business Cycles, Indicators, and Forecasting (NBER Studies in Business Cycles, Vol.28), pp 95-156.

Sykes LR, Jaumé S (1990) Seismic activity on neighboring faults as a long-term precursor to large earthquakes in the San Francisco Bay area. Nature 348: 595-599.

Tukey JW (1977) Exploratory Data Analysis. Addison-Wesley Series in Behavioral Science: Quantitative Methods. Addison-Wesley, Reading (Mass).

Turcotte DL (1997) Fractals and Chaos in Geology and Geophysics. 2nd Ed., Cambridge University Press.

Turcotte DL (1999) Seismicity and self-organized criticality. Phys Earth Planet Int 111: 275-294.

Turcotte DL, Newman WI, Gabrielov A (2000) A statistical physics approach to earthquakes. In: Geocomplexity and the Physics of Earthquakes, Am Geophys Un, Washington DC.

USDL (1999) U.S. Department of Labor, Bureau of Labor Statistics, Labor Force Statistics from the Current Population

Survey.   Web   site:
http://stats.bls.gov/webapps/legacy/cpsatab5.htm.

Varnes DJ (1989) Predicting earthquakes by analyzing accelerating precursory seismic activity. Pure Appl Geophys 130: 661-686.

Vil'kovich EV, Shnirman MG (1983) Epicenter migration waves: Examples and models. In Keilis-Borok VI, Levshin AL (eds) Mathematical Models of the Earth's Structure and the Earthquake Prediction, Comput Seismol 14, Allerton Press, New York, pp 27-36.

Vorobieva IA, Levshina TA (1994) Prediction of a second large earthquake based on aftershock sequence. In: Chowdhury DK (ed) Computational Seismology and Geodynamics, Vol 2, Am Geophys Un, Washington, DC, pp 27-36.

Vorobieva IA (1999) Prediction of a subsequent large earthquake. Phys Earth Planet Inter, 111(3-4): 197-206.

Vorobieva IA, Soloviev AA (2005) Long-range interaction between synthetic earthquakes in the block model of lithosphere dynamics. In: Chowdhury DK (ed) Computational Seismology and Geodynamics, Vol 7, Am Geophys Un, Washington, DC, pp 170-177.

Wyss M, Habermann R (1988) Precursory seismic quiescence. Pure Appl Geophys 126: 319-332.

Yamashita T, Knopoff L (1992) Model for intermediate-term precursory clustering of earthquakes. J Geophys Res 97: 19873-19879.

Zaliapin I, Keilis-Borok V, Axen G (2002) Premonitory spreading of seismicity over the faults' network in southern California: Precursor Accord. J Geophys Res 107(B10): ESE5-1 – ESE5-15, 2221, doi:10.1029/2000JB000034.

Zaliapin I, Keilis-Borok V, Ghil M (2003a) A Boolean delay model of colliding cascades. I: Multiple seismic regimes. J Stat Phys 111(3-4): 815-837.

Zaliapin I, Keilis-Borok V, Ghil M (2003b) A Boolean delay model of colliding cascades. II: Prediction of critical transitions. J Stat Phys 111(3-4): 839–861.

Zoller G, Hainzl S, Kurths J (2001) Observation of growing correlation length as an indicator for critical point behavior prior to large earthquakes. J Geophys Res 106: 2167-2176.

www.ingramcontent.com/pod-product-compliance
Lightning Source LLC
Chambersburg PA
CBHW080757300326
41914CB00055B/929